# THE ART OF FIRE
## THE JOY OF TINDER, SPARK AND EMBER

CENTURY

# CENTURY

20 Vauxhall Bridge Road
London SW1V 2SA

Century is part of the Penguin Random House group of companies
whose addresses can be found at global.penguinrandomhouse.com.

Penguin
Random House
UK

First published in 2017 by Century

www.penguin.co.uk

A CIP catalogue record for this book is available from the British Library.

ISBN 9781780897660

Designed by Tim Barnes ⟨⟩ herechickychicky.com

Typeset in Givens Antiqua and Camphor 8½/15pt by Monotype

Printed and Bound in Italy by L.E.G.O. S.p.a.

Penguin Random House is committed to a sustainable future for our business, our readers
and our planet. This book is made from Forest Stewardship Council® certified paper.

MIX
Paper from
responsible sources
FSC® C018179

# THE ART OF
# FIRE

## THE JOY OF TINDER, SPARK AND EMBER

# DANIEL HUME

### ILLUSTRATIONS BY
## ADAM DOUGHTY

# CONTENTS

## INTRODUCTION

Fire is mankind's oldest energy.
It must've left a love of fire inside human genes.
**Lars Mytting**

Have you ever gazed into the flickering red and orange flames of a fire and felt mesmerised, inspired and energised? Fire is at the root of the progress of the human race, and the instinct to huddle closely around, facing inward towards the flames warming our hands and faces, is one of the most ancient and deeply rooted of all. Fire fascinates, captures the imagination, and brings families and communities together. Like any of the wonders and mysteries of the universe, it has the ability to touch something deep inside all of us.

For hundreds of thousands of years, at every corner of the globe, humans have relied on fire. From the bushmen in the Kalahari desert, rhythmically circling their fire as part of a ritual trance dance to heal a sick child, to Sami reindeer herders in northern Finland, waiting patiently for their kettles to bubble over the flames, fire brings its benefits to us all.

Fire plays a central role for all of us. It provides the basic and primitive essentials of light, heat, energy and cooking, and extends far beyond this to possess unique spiritual and cultural significance for many. Fire connects us to others, our emotions and our history. Our memories are imprinted with it, and it helps to define our communities and our lives.

My memories of fire are as clear as glass: distinct snapshots like opening much-wanted gifts on Christmas mornings and birthdays. When I was barely as tall as the fireplace itself, I recall helping my dad screw up balls of old ink-smudged newspaper and placing them neatly in the grate before carefully arranging split pine kindling over the top. We lived in an old Victorian house with high ceilings and draughty corners, and there was always an open fire in the living room during the colder months. We would mostly burn coal, but would sometimes throw an old log on when we had wood that we needed to get rid of. The matches and lighters were kept in an old, rusty-edged blue tin placed on the highest surface in the kitchen, well out of the reach of my and my brothers' small, fearless hands.

I am the eldest of three boys. My brothers Ben and Sam and I grew up in the countryside in the Stour Valley, on the Essex and Suffolk border. Like many children in the area, we spent most of our time roaming around through nearby fields and woodland, getting caked from head to toe in mud. After having a bath, I would stand in front of the fire with a towel draped around me until I was completely dry and warm. Gradually, as I got older, my parents gave me more responsibility when it came to the fires at home. Soon I was lighting the fire in the living room on my own, taking a chair over to reach for that blue tin so I could strike a match to set the fire ablaze. On other occasions, we would light huge bonfires in the garden. My dad would sometimes make a torch for us by taking a short branch from the pile of wood, wrapping a piece of old hessian sacking around the end, pouring vegetable oil over it and setting it alight before handing it to me to gaze at in awe. My brothers and I had fun, but were taught the importance of respect, especially when it came to the immense power of fire.

As we grew older, our free time took an altogether different tack as we continued to play and explore outside. I dreamed of the wilderness day and night, and learned anything that could help take me there. This area had been our family's home for at least 350 years, and I thought about barely anything else except walking the same outdoor paths as my relatives and ancestors, discovering as much as I could about the natural world. As a child, I devoured every piece of literature I could find relating to bushcraft and fire-making. The ability to accurately and quickly light a fire is one of the most important skills anyone heading off on a wilderness adventure could possess, yet very little is written about it in a complete way, based on first-hand experience.

At seventeen, I got a job at Woodlore Limited, the UK's leading school of wilderness bushcraft and tracking, a role I enjoyed until stepping down as Head of Operations in 2017. I continue to teach others. I have devoted my life to learning as much as I can about the natural world, and I have travelled across the globe in a quest to find out more about fire's place in our lives and the communities that use it.

Throughout this book, I will frame practical step-by-step explanations of the different ways fires can be conjured, drawing on my experiences, during my travels and teaching, of making fire and of the traditions involving it. I will also cover wider topics related to fire, such as its place in history, culture and spirituality, and ultimately how it shapes the world around us.

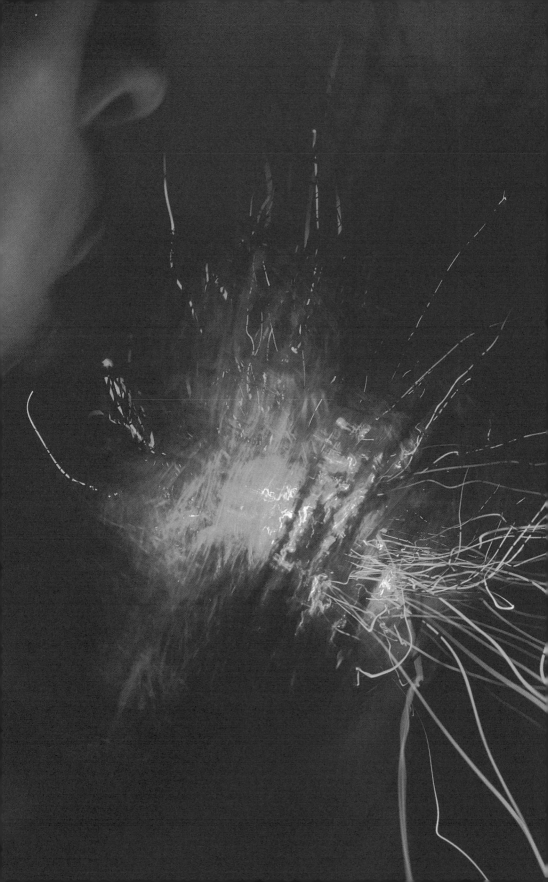

CHAPTER 1

# EMBERS & TINDERS

**WHEN I WAS EIGHT, I LEARNT IT WAS POSSIBLE TO MAKE FIRE BY** friction during a conversation in the playground at primary school. One of my friends told me that his dad said you could make a flame by rubbing two pieces of wood together. I was sold; I needed to find out more. The *SAS Survival Handbook* by John Wiseman subsequently became my preferred bedtime reading and was soon well-thumbed, with bits of mud sticking the pages together. I began experimenting at home with any piece of wood I found lying around the garden – including a piece of fence panel, which I pressed into action as a crude fire plough. It will come as no surprise to hear that I had little success with this.

By the time I was ten, I was lighting bonfires of my own in the garden. I became very proficient at starting one in the fireplace indoors, lighting it in only a minute or two. I learnt how to manage a fire and was fascinated by it, watching it burn and how it reacted to more fuel being placed on it. I started to cook on fires, making stoves from bricks in the garden. I loved sleeping outside under the stars next to my smouldering bonfire, the dying embers glinting in the moonlight.

From there, I turned my attention to other skills relating to bush-craft, still relentlessly trying to make a fire using friction. I began by using the hand- and bow-drill methods and would attempt to make a fire every day after school, my school bag flung hastily by the door as I headed for the back garden. I got quite good at using the hand-drill technique, although success was slow, as I was unknowingly using a very difficult wood. The bow drill was different, though: hours, days, weeks and months went by, and still I felt I was no closer to lighting a fire. I was emulating what I had read in survival manuals to the very millimetre, but still progress was painfully laboured.

While my face contorted with frustration, an internal determination kept me from giving up. I pushed on despite the apparent overwhelming odds. I lost count of how many times I had skinned my knuckles on the patio or worn out the various wooden parts of the bow drill I was using, forcing me to search for another piece of wood and carve new parts again and again. Despite these problems, thoughts of defeat never entered my brain; they were kept at bay by any hints of progress. I really believed I could do it. Every time I practised, I noticed small improvements, and occasionally a big leap would spur me on even more.

I had just turned fourteen when one day my practice paid off at last. I was on the lawn outside the house, going through the same motions that I had done countless times before. I had switched my bow for the one from my bow and arrow set because it was longer, and I felt I would be able to get a better spinning action. I bowed as hard as I could, watching the hot, black dust being ground off by the friction of the two wooden parts of the set. I saw it collect into a small pile. I had been through this process literally thousands of times but this time, something felt right. I threw down my bow and peered at the little pile of charred dust that had accumulated. It continued to smoke on its own, like a cigar, even though I had ceased drilling. Finally, an ember. I had done it!

Overwhelmed with the feeling of success, my hands trembled with delight and fatigue. But it wasn't over; I was only halfway. I needed to feed this fragile little ember and transform it into flames or it would die. I waited for a few crucial moments to allow the ember to grow from the centre of the pile until an orange glow was visible, like the crater of a miniature volcano. I sped up the process by directing the merest whisper of a breath onto it, being careful not to obliterate it. Imagine the tip of a cigar when someone is drawing on it; that was the glow I was waiting for. And there it was: my cue to transfer it to some tinder.

My heart sped up as it dawned on me that I had no tinder in my pocket. Everything around me in the garden was damp. My thoughts racing as quickly as my pulse, I ran over to our rabbit hutch and frantically grabbed a handful of straw from the bed of little Leo, the family's much-loved Lionhead rabbit. Back at the ember, tinder bundle in hand, I felt relief to see a wisp of smoke still emanating from it. I pinched the little shaving of wood that the ember sat on, lifting it up off the ground. I placed it inside the bundle of straw and blew on it gently. Each time my carefully directed breaths met the ember it lit up, and the glow became intense as it shared its infectious heat with the dry tinder around it. It grew and grew over a period of thirty seconds or so, and then I heard it roar. A bright flame engulfed the bundle as well as my fingers, and as I released my grasp it dropped onto the ground. The flames devoured the bundle and singed the lawn. I jumped up and down with delight and ran indoors to tell everyone, leaving a smouldering pile of blackened ashy remains. Finally, I had lit a fire not only on the ground but also deep inside myself. Today, long after that ash has scattered away in the breeze, the internal flames continue.

## EMBERS

In this book, I will describe a vast array of ways to start a fire and manage it to your needs. The way in which we produce flames is dictated by the method we employ to create the initial source of heat. These methods can be split into several categories. Some produce flames directly, such as modern sparking devices and matches, while others only produce an ember. Broadly speaking, embers tend to be the result of the older techniques, and require gentle and careful coaxing, along with the addition of fine tinder.

## TINDER

The term 'tinder' is a large umbrella under which all sorts of materials may be included, but they all have a common purpose: they capture, sustain, and transfer that first tiny quantity of heat, whether it comes in the form of a spark or an ember, to the next stage of fuel. The type of tinder should be selected according to the particular method chosen to create that initial heat source. Most methods of ignition produce such tiny, short-lived quantities of heat that the tinder has to be absolutely perfect – meaning it must be completely dry – and some methods dictate that the tinder must also be fibrous and fine.

Once you have a sustained fire that is burning fiercely, you can sometimes get away with having main fuel that is less than perfect, but with tinder, because the initial heat source is usually so insignificant, it really needs to be faultless. If your tinder is not very dry, you will have won half the battle only to fall at the home straight.

Experienced wilderness travellers will always be vigilant to available tinder as they move through the country. They have an opportunistic attitude, keeping in mind the medieval proverb, 'Make hay while the sun shines.' If they come across tinder somewhere along their journey, they will stop and collect it, filling a jacket pocket or stuffing a load into a spare space in a rucksack. By contrast, an inexperienced traveller will arrive at a camping spot in the rain, later than anticipated, and as time begins to speed up and dusk approaches with alarming swiftness, their minds will turn to a warming cup of hot tea and some supper. Travelling lightly without a gas stove and reliant upon fire to boil the kettle, they will probably realise that the forest around them is completely soaked, as are their hands.

There are ways to light a fire even in the worst weather and in the most ill-prepared of scenarios, but it makes life easier, safer and vastly more enjoyable to be ahead of the game. The advice I was given when I was learning about bushcraft – and the advice I give now – is that if you find yourself accompanying an experienced person on your travels, be sure to keep a sharp eye on what they do and in which order they do it. There is an intricacy, a subtlety, an understated nuance to their actions which is very easy to miss or overlook. If carefully observed, though, and coupled with practice and your own experiences, it can enhance your understanding immeasurably, and in a way that simply reading or sitting in a classroom never could.

The greater the need for a fire, the more difficult it is to create. Holes or chinks in one's outdoor ability never fail to show themselves; nature has a clever way of exposing them, and none will catch the inexperienced or complacent out more than the skills relating to fire. The key to fire-starting in any scenario is attention to the tiniest of details.

When searching for tinder, bear in mind that you do not neces-
sarily need to know the name of the plant or tree species, or even recognise it, in order to decide if it provides a suitable material for tinder. Of
course, it can pay to learn about the species wherever you are venturing
off to, because some can be harmful, particularly in the tropics. That aside,
though, the principle is very simple: generally, any material which is dry,
fibrous, quick to gather and is easily available in the necessary quantities
can be used to start a fire.

Some tinders are restricted to one method, while others are a 'jack
of all trades' and can be applied to a variety of approaches. Wherever in
the world you encounter traditional fire-making, the tinder used is almost

ABOVE A tinder bundle composed of palm leaves and bamboo scrapings collected by the Semai people, state of Panang, Malaysia.

exclusively of a vegetal substance; the use of animal substances is less common. That being said, there are interesting accounts of various indigenous people gathering the downy feathers naturally shed from birds. One account even records use of the felt-like nest lining spun by a South American species of ant called *Polyrhachis bispinosus*.

Tinder is also not restricted to naturally occurring products found in the environment, and there are many modern materials that can be used. It is usual to employ natural products when we are travelling, because there is often an endless supply, and it is best to save any materials we take into the bush with us and use them only if necessary. For example, it is wise to carry a couple of chemical firelighters tucked in your equipment somewhere; they may never come out of your pack, but they are there in case you need them. The list of tinders that follows is by no means exhaustive, but it suggests a few species that are suitable. When I am teaching, I always advise my students that they familiarise themselves with some of them, then take those principles further afield and apply it to their own travels, wherever that may be.

### BIRCH BARK

The papery, multiple-layered outer skin of this tree contains the flammable substance betulin and can be removed from the trunk in whatever size sheet is needed. When using a knife to gather bark from trees, always try to collect from those that are dead and fallen and not from those still standing. It is unsightly when you come across a tree that has been de-barked in this way – and worse, if done incorrectly, it can kill the tree. The bark from standing trees will still work well, though, and indigenous people will often gather it in this way, usually to make containers or other objects requiring greater suppleness than is found with the bark from a dead tree.

Quite often you will see the bark naturally shedding from standing trees, and it is fine to go along and pull it off, being very careful not to remove any more than is peeling naturally.

If you cannot find any naturally peeling bark, find a fallen tree and score along its length with the tip of your knife, cutting down to the woody inner bark. Having done this, you can prise it off with your fingers or, if the bark is being stubborn, carve a suitably sized dead stick into a flathead screwdriver shape – a 'spud' – and use this to help remove it. In Britain, or anywhere else with a similarly mild climate, this will not require you to cut very deep. As you travel further north and into places with colder winters, however, you will notice the bark has more layers to it and can be surprisingly thick – and the thicker the better. It is also worth noting that although the combustion will be slightly impaired, birch bark is rather unusual in that even after a good soaking it will still burn, as long as the majority of water is shaken from the outside and it is patted dry on a trouser leg.

Birch bark can be lit in two main ways. The quickest is directly, using a match or other direct flame, the bark requiring no preparation; the flame is simply offered to the edge of the bark. This will soon turn into a furious blaze. If a direct flame is not available, it can be lit easily with sparks too. To accomplish this, the bark must be prepared by scraping the pale-coloured outer side with a sharp edge. The aim of this is to produce a fibrous cluster of fine, dry scrapings about the size of a golf ball. A spark from a ferrocerium rod can then be dropped into this cluster. If done correctly, the scrapings will immediately burst into flame. Once birch bark is burning strongly, it is quite difficult to extinguish, and its combustion is not too dissimilar in vigour to a chemical firelighter we might use at home – even the smoke given off is black.

Before the flames take hold properly, though, it must be handled carefully, otherwise it can be accidentally extinguished. Birch bark has a tendency to curl up tightly and snuff itself out as soon as it gets hot. To avoid this, you can either fold it, concertina fashion, in line with the grain, so as to form a fan-like sheet, or you can tear off lots of long, thin strips and screw them up into a mass about the size of a tennis ball. These two methods work well if you are applying a direct flame, but are inconvenient if you are using sparks. If this is the case, be ready both to hold the bark open to prevent curling, and to add the thinnest strips to the initial flame.

## FIBROUS BARK

There are countless species of plant and tree that provide suitable bark. They can be split into two main categories: those with suitable outer bark, and those with suitable inner bark. Neither kind is difficult to gather. Whatever bark you gather, it must be dry and as fibrous as possible. Usually this means initially splitting it with your fingers before rubbing it vigorously between your hands. This will make the bark a lot finer, and will allow it to burn more easily and readily.

### OUTER BARK

Outer bark is more obvious than inner bark, being visible without the need to stop and check, as is usually necessary with the species that provide inner barks. You will often see it hanging down in ribbons or at least separated from the main stem, the plant having shed it naturally. Rather like the natural peelings from the birch tree, this bark can be gathered quickly. Providing it has not rained recently, it will have been air-dried and will require very little (or more likely no) drying before use. Some species will not be so obvious as this; nonetheless, you should still see a fibrous exterior to the bark which will indicate it might be of use and should warrant further inspection. Examples of species that provide outer barks are honeysuckle, clematis, willowherb, red cedar and juniper.

### INNER BARK

As the name suggests, inner bark is located under the covering of the outer bark. Some species provide a very fibrous, stringy material that can be gathered quickly. Search for fallen trees that are beginning to lose their outer bark to decay. Pull the outer bark off and look underneath. If you find a good example, you will be able to pull off decent-sized sections and stringy lengths of dead, dry bark, although some species and trees in less than ideal condition can be a little fiddlier and time-consuming to gather from. This bark

LEFT A woman blows an ember into life in tinder made from shredded leaves
and the cloth-like material from the trunk of a palm, New Ireland Province, Papua New Guinea.

can also often be damp, but providing it is not raining, there only needs to be a slight breeze and it will dry out fast if hung up in lengths in the trees. It can also be put in a pocket to allow your body heat to dry it, but I have found this less effective. If you have good weather, it is better to hang it up to dry. Examples of good sources of inner barks are sweet chestnut, poplar, lime, oak and coconut husk.

### DRY GRASSES

With over 11,000 accepted species of grass providing approximately 20 per cent of the vegetation cover on earth, dry grass is very commonly used as tinder all around the world, and it works very well. The grass you gather must be dead and dry, and the best grasses are those with thin, wide leaves rather than round ones. This is because there is more surface area and that helps the combustion. It is very quick and easy to gather – comb out the dead leaves from the living ones with your fingers.

### DRY LEAVES

When you imagine dry leaves, you tend to think of those dropped from broad-leafed trees. However, these leaves tend not to be so good as tinder, very often smouldering and not producing a lasting flame without considerable effort and perseverance. They are usually also found on the ground, which means they will have been sitting in dampness.

The best leaves are those from the previous season's ferns. Luckily, ferns are a large family and quite widespread. When collecting them to use with sparks or small embers, be strict and gather only the finest leaves, and the thinnest stems. Do not let the thicker stems enter your tinder bundle, as their thickness will hinder your fire-making attempts. If the ferns are wet from morning dew or rain, the best way to dry them if the weather improves is to leave them and let the breeze do the work. It is tempting to collect damp leaves and put them somewhere else to dry, either laid out on the ground or in a pocket, but if you leave them where they are, you will find they will dry much quicker. Be careful when collecting fern leaves as the stems are very sharp if they become crushed, and I have seen some deep cuts on the hands as a result.

Kapok seed head from Komodo Island, Indonesia.

## PLANT DOWN

The downy seed heads from various plants can be used as tinder. Examples include clematis, willowherb, goat willow catkins and cotton grass. In the tropics, some of the best tinders are the cotton-like fluff from the seed pods of the kapok tree, also known as the Java cotton – its large pods festoon the tree and give off large quantities of tinder material. It can be picked directly from the tree if within reach or gathered off the ground in dry conditions. Even better than this is the fluffy scurf that can be collected directly from the trunks of fishtail palms.

## DRY WOOD SCRAPINGS

A little-known and underrated method of producing tinder in difficult conditions is by using dry wood scrapings. Sometimes you will not be able to find any other dry tinder; if the weather is bad, in particular, it may be best to split open a dead piece of wood or carve off the damp outside, and scrape the dry wood inside with a sharp edge. Hold the blade with both hands to achieve maximum support and steadiness, and proceed to scrape at 90 degrees to the wood's surface. It is quite a time-consuming task, but give yourself the best chance by ensuring you have a nice smooth surface to scrape. Again, make sure you hold the blade with both hands. You will swiftly build up a sufficient quantity of scrapings to start your fire.

### DRY, DECAYING WOOD

Also known as 'punk', decaying wood can usually be found in a dry state if you search in the right areas. Look in hollows, and in nooks and crannies in trees. Old stumps also provide it, but as it is likely to be more exposed it will probably be saturated most of the time, depending on the environment you are in. You will know when you have found some because it will be very light in weight and will crumble easily in the hands. The best stuff will catch a spark and continue to smoulder of its own accord until it has completely consumed itself.

FUNGI

Many species of bracket fungi (those which grow on wood) can be used for tinder. They have been tremendously important throughout history and continue to be of use today. However, the only time you might need to make use of them is in a survival situation in which it is only possible to produce sparks by striking a flint or similar stone with a piece of carbon steel or iron pyrites (see Chapter 8). This is because the sparks produced from these methods are cooler than those made by modern devices, and therefore the range of tinders that can be used is more limited. Fungi are also very good tinders for use with the solar methods of ignition such as a magnifying glass. This is due to their dark colour and their ability to smoulder without much encouragement.

## HORSE HOOF FUNGUS
### (Fomes fomentarius)

Probably the most well-known fungus is commonly known as horse hoof fungus, which grows widely across Asia, Europe and North America. Although it can occur on several species of tree, it is to be found predominantly on birch. In mainland Europe, 5,000-year-old Ötzi the iceman, the well-preserved natural mummy found in 1991, was found to be carrying small pieces of this fungus threaded onto a string inside a small leather pouch, which also contained flint and remnants of iron pyrites for making sparks. It was the fire-lighting kit of the day, and remained so right up until the advent of steel, which replaced iron pyrites because it produced a more versatile spark. A tinder called 'amadou' was produced from this fungus, and its use for fire-lighting was taken advantage of until quite recently; in fact, Germany even produced amadou matches with striking heads into the twentieth century.

There are two main ways this fungus can be used. Firstly, it can be used to carry fire from one place to another. For this purpose, it can usually be put to use straight off the tree with no preparation required. This would have been of tremendous use for our ancestors, allowing them to move from one place to another without the need to start a new fire from scratch, thus saving valuable materials and time. A dry specimen is removed from the tree, and the underside placed into a fire for a couple of minutes before being removed and encouraged with a few breaths of air. The fungus can be left to smoulder very slowly with no flame at all. Depending on its size, it may burn for several hours as it consumes itself.

Horse hoof fungus growing on a fallen branch.

Exposed trama layer.

Secondly, it can be used to start a fire from scratch, which requires preparation. The hard, horny exterior on the top can be carved away with a sharp knife to expose a soft, brown layer called the 'trama', which is not dissimilar to suede leather. Once exposed, it is possible to carve off broad, thin sheets of this material rather like one might carve a chicken breast, aiming for the widest, thinnest pieces. Taking one of these pieces in the hands, it is manipulated and pulled gently apart, with the aim being to stretch the sheet. As you do this, you will see and feel the fibrous structure of the fungus sliding apart.

The sheet will grow, and sometimes even double in size. Occasionally you will go too far, and a little hole will form, but do your best to avoid this. The material will now be akin to the softest and most beautiful suede leather you have ever felt. You will want a jacket made of the stuff! In fact, all sorts of items are made from this material in Eastern Europe even today. All that is sometimes required is to dry the sheets out if they are damp.

## KING ALFRED'S CAKES
### (Daldinia concentrica)

King Alfred's cakes is a hard, black fungus that grows in groups, mainly on dead ash wood. It looks like a charcoal briquette and burns rather like one too. Found mainly in areas of North and South America as well as most of Europe, it can be prised off the tree and used immediately if dry. You can tell if a specimen is wet inside by feeling its weight; light means dry, heavy means full of water. The colour can also provide a clue – the black ones are the most likely to be dry as opposed to brown specimens. Having broken one open to expose the silvery rings inside the fungus, it will catch a spark and smoulder on its own until it has consumed itself, requiring no tending. Although no archaeological evidence has yet been discovered, I have no doubt that this fungus was also used to harness the sparks from iron pyrites and flint in areas or at times in which the horse hoof fungus was unavailable.

**CHAGA**

*(Inonotus obliquus)*

Occurring most commonly in the great boreal forests of northern latitudes, chaga (ABOVE) can also be found further south in some parts of Europe. Preferring birch as its host tree, this fungus with its fissured black exterior looks like the tree's equivalent of a tumour, and stands out against the pale colour of the birch bark from which it erupts. Some can grow to the size of a man's head or larger. A piece of this exterior can be prised away or chopped off with a hatchet to expose a rich golden-brown interior, which will catch a spark and smoulder slowly, providing it is absolutely dry. The Cree, an indigenous group from North America, called this fungus 'pesogan'; it is sometimes also referred to as 'touchwood', and as well as fire-lighting, the Cree use it in ceremonies and as a treatment for arthritis.

### HERBIVORE DUNG

In arid regions, old herbivore dung can be found in a dry state which lends itself to fire-lighting. It is usually made up of digested vegetable matter and is therefore very fibrous. In some areas where wood is hard to come by, people even burn dung as a main fuel. In Africa, elephant or rhino dung may be used, and in Australia, kangaroo dung.

The finest 1mm-thick twigs around the lowest branches of some coniferous trees make very convenient tinder because they are usually dry, having been sheltered from rain by the rest of the foliage. A bundle of the thinnest twigs can be lit with a ferrocerium rod if nothing else is available, but more usually they are lit with a flame from a match or lighter, and get a fierce blaze going very quickly.

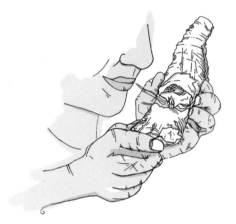

## FIRE DOGS

When you come across an old campfire site, even one that is a few months old, or if you need to re-light your own, having woken to find it burned out, it can be done very easily by making use of any leftover charcoal and half-burned logs – known as 'fire dogs'. Sparks, or preferably a single lit match, can be dropped onto the charred surface of a log, and providing the charred materials are dry, will continue to smoulder. The best place to drop sparks is in an area that has tiny flecks of white on the charred surface. As soon as you see the smoulder, put your fire-lighting equipment safely away, and begin to blow onto the smouldering area, encouraging it to spread until a wider area of wood has caught. Take another similarly charred log and hold it alongside the burning one so the two touch each other. The ember will spread onto the second log. Add a couple more logs in the same manner, and continue to blow. Once you have three or four together, a flame will soon erupt from the pile and will no longer require breaths. Most woods will work, but willow is particularly good.

In the jungle or other damp environments, it can be very useful to carry some strips of old inner tube in your shirt pocket. The beauty of this is it does not absorb water, will light very easily from a flame, and burns hot and long enough to light kindling. In an emergency, you could even cut small strips from the soles of your boots.

## CANDLES

In the wilderness, we sometimes carry candles with us, because they have more uses than lighting our camps. They are great for lighting fires, being longer lasting than a match. If needed, the end of the wick can be teased apart a little and lit with sparks from a ferrocerium rod (see page 145).

## LICHEN

The most useful of lichens are the beard lichens that hang from the branches of trees. The *Bryoria*, *Alectoria* and *Usnea* genera include the best. Quite often lichen are damp, and therefore need to be dried out before use. They can be found dry, though, so search thoroughly in sheltered spots. Lichen also survive well in extremely dry and cold environments. In these instances, they can often be found in a dry state and used immediately. In very cold temperatures, it does help to warm any tinder up in your pocket for a minute or two before you attempt to light it. In winter, when the ground is covered in snow, and in areas without birch trees, lichen is sometimes the quickest way to light a fire.

CHAPTER 2

# HAND DRILL

**EVERYTHING ABOUT THE DELICATE BALANCE OF LIFE AND OUR PLACE** within the natural world fascinates me. I enjoyed primary school, but at secondary school, I started to lose interest in lessons, my eyes glazing over as I daydreamed about my next adventure. Homework sat untouched in my school bag, and I spent increasing amounts of time on my own in the woods and fields near where I lived, honing my skills and knowledge. My parents agreed to let me leave school when I was fourteen, and my mum home-schooled me. I eventually sat only two GCSEs (in English and maths). Academic achievement was of little importance to me; I knew I belonged in the Great Outdoors.

My favourite textbook back then was Ray Mears's *Essential Bushcraft*. A dog-eared copy sat proudly on my bookshelf and often travelled with me. Ray was someone I greatly admired, and I had read about Woodlore, the UK's premier school of wilderness bushcraft, which Ray founded in 1983 to teach small groups of students the knowledge and techniques that he had spent many years honing. I was very keen to go on one of his 'Fundamental' courses, which taught the basics of bushcraft skills and knowledge. My parents booked me onto the junior course, but at one of Ray's book signings I persuaded the lady who managed the bookings to bump me onto the six-day course for adults, despite only being sixteen.

I enjoyed the course immensely. I had already practised a lot of the skills we went over and yet I still learnt a lot. More importantly, I had finally found other people who spoke my language. These were people who, like me, felt truly at home in the countryside, amongst the trees. I loved being in the outdoors under the stars, warming my palms on flames that I had conjured.

The next course I booked myself onto was learning how to success-fully track a human or an animal in the forest. When I applied, I also asked – with the eagerness of an energetic puppy – to do some work experience.

They agreed, and after assisting Ray and the other instructors, I was offered a part-time job. I was seventeen. Opening the letter and reading that someone was going to pay me to do something I loved felt like all my birthdays had come at once.

A year later, we were to go on a seventeen-day course in Namibia led by Ray and another instructor, Bob, an old friend of Ray's who had grown up in Kenya. I had dreamt of Africa: the endless vistas, the stunning wildlife, the varied human cultures, and the survival techniques many of them used. I knew I had to start saving as much as I could to contribute towards the cost of the trip, so worked hard at the local butcher. All the while, in my mind I was going over the things I was most excited to see, such as how the indigenous people used the hand drill to make fire.

We travelled to Windhoek, the capital of Namibia, on a nine-hour direct flight from Heathrow. It was over 40°C when I landed and walked down the metal steps onto the dry earth below. A hot wind blew across the airstrip, almost taking my breath away – it was like being on the receiving end of a giant hairdryer. We headed north-west to a place called Hobatere, where we learnt the art of tracking. After a few days, we continued further north to Etosha National Park and to the area formerly known as 'Bushmanland', looking for a small settlement called Tsumkwe. It is a long, long way from anywhere. The roads turn into gravel, and then to dust, for hundreds of kilometres. It is a bushman community hub that has many satellite groups dotted around it. From Tsumkwe, we headed a further 15km into the bush along dusty, dirt tracks, where we arrived at a San community of around twenty-five to thirty local people.

The San people once occupied all of southern Africa, but are now restricted to small isolated pockets in Angola, Botswana, Lesotho, Namibia, Zambia and Zimbabwe. They are thought to be the oldest race on earth. They are the ultimate minimalists and complete masters of travelling light. They have to be: this is not an environment in which to be laden with clutter and heavy gear. Their usual attire is a loincloth made from antelope leather and decorated with beaded patterns. Some wear an old jumper in the cooler evenings, as it can drop below freezing at some times of the year. Materialism does not exist in their culture, and there is a lack of emphasis on prestige and wealth. They have a sad history of poverty, decline in their cultural identity and discrimination. Yet they have also received a lot of attention from anthropologists due to their survival and hunting skills, and their wealth of knowledge about the local flora and fauna. They have proud and intuitive ties that bind them to the land; their lives blend the natural world and spirituality amid harsh, arid environments.

The settlement was in a clearing, with a dozen grass and reed huts in a semi-circle. Each family occupied a small hut with a fire outside, and there was a central fire in the middle of the clearing, around which the community congregated in the evenings. The fires were used for cooking and light, but also to stop hyenas and leopards coming into the camp. The men live for hunting; a lazy hunter is considered a source of shame to the entire clan. They carried bows and grass-stemmed poison arrows. Their prey, which includes antelope, giraffes and zebras, dies not from haemorrhaging but rather from a pinprick of potent poison, squeezed from the inside of a leaf beetle larva onto the arrow shaft. They are not wasteful: the meat would be boiled or roasted on the fire, animal hides tanned for hunting bags and blankets, and bones cracked for marrow. On this trip, there were a few bizarre reminders of the modern world; some men carried quivers made from drainpipes, and there was litter strewn around. I learned that many families moved to Tsumkwe so they could send their sons and daughters to school, and some men went to work early as labourers on farms, returning in the evenings or at weekends. However, they were still very traditional, and many of the older men only spoke their ancient language – a combination of delicate sucking married with clicks.

The tribe had cleared a huge area of ground using their bush knives for us to pitch our tents in. In exchange, we helped them gather mangetti nuts from the mangetti tree; we eventually filled huge sackfuls of them.

OPPOSITE San bushman with hand drill set and quiver containing arrows and spare drills, Namibia.

Foraging for food was mainly the women's job. They have knowledge of over 300 different species of plants, many of which have medicinal purposes, from combating hunger to treating rheumatism. They showed us how they carry digging sticks and took us to dig up tubers which looked like giant potatoes, the flesh of which they squeezed for a refreshing drink. They also showed us how they made jewellery using cut-up polished ostrich eggshells; some of them sold these on to tourists to make extra money.

The men took us out hunting and to gather sticks to use for the hand drill. They would walk out in single file through the bush to collect mangetti wood. The mangetti trees were huge, with thin straight suckers at the base. We took the sticks back to camp, where the San people showed us how they made fire. There are always variations within different fire-making techniques. For the hand drill, I had always thought that the bottom piece of wood – the hearth – was wide and flat. The San, however, used old drills as hearths, so used two sticks of the same thickness. Watching them work was exhilarating. Their drills had amazing spiral patterns charred on them, which looked incredible when the drill was spun. Although I was not sure exactly how they made these patterns, I suspect they wrapped a piece of damp cloth or bark around the drill and put it in the fire. We communicated by gentle hand signals and gestures, wry laughs and wide smiles. I felt truly privileged to observe the way they live.

On the final night, we witnessed a 'trance dance'. I had heard about this spectacle, and had noticed the well-trodden circular path around the fire when we had first arrived. The San believe that the trance dance has a direct communication with the spirit world, and that their souls journey through space and time to connect them to spirits and to their ancestors, who will give them insights into survival, such as wisdom and healing powers. Many dances are held as curing dances, with an aim to improve the health of one of the members of their community and to draw out the 'arrows of sickness'. The men – the dancers – circle, while the women's singing and clapping brings them to a frenzy, as the spirits rise upwards and cause them to collapse – known as a 'half death'. The dancers move out of their bodies, where they meet with the spirits who give them power. When they return to themselves, they channel the power to the people around the fire by touching their shoulders. It is a social event, an opportunity for self-expression. It goes on for hours, and ends before the sun fully rises.

As we watched, the men circled the women with small, persistent heavy footfalls in the sand around the fire, their bodies hunched forwards with arms close to their sides, their figures tightly flexed. They wore rattles around their legs made from cocoons filled with small stones, which jangled loudly and incessantly. The women sang noisily and methodically, some low and loud, others more softly in a higher pitch. This ancient song had been passed down through thousands of years – endless improvisation within the bounds of repeated musical phrases. The communal, insistent clapping rhythm reverberated in my ears. Rather than connecting with the spirits – this was done in private – this was a sociable and joyous occasion. It felt primal, energising and uplifting. Music is important to all cultures as a form of expression. Like fire, it connects us to other human beings, our emotions and our history in a way words alone cannot. It helps to define our lives, our communities and our prayers.

Later, we sat around the fire and handed the grinning children balloons and the adults T-shirts. As we warmed ourselves as the temperatures dropped, eating food our hosts had killed and cooked, chatting and gesticulating, it was a necessary reminder that fundamentally, we are all the same.

**HAND DRILL SET** from West Papua Province, Indonesia.
Note the extra-small notches.

# USE OF THE HAND DRILL

The hand drill is made up of two sticks: one is a drilling stick, long and thin, and the other is a hearth stick, which is usually flatter and wider than the drill. The drill is held vertically and twirled rapidly between the palms of the hands, while the bottom end is seated into a slight depression in the hearth. A small notch is usually cut into this depression in order to give the friction dust somewhere to collect and form itself into an ember. The hand drill once had the widest distribution of all the friction fire-making techniques, being found all over the world. Today, it has fallen out of use; there are people in remote areas that still rely on it to make fire, but this is becoming rare.

The hand drill's range is loosely restricted to the warmer, drier regions. In colder, damper environments, the method becomes less reliable because these fire-suppressing elements can both impregnate the vulnerable wooden parts during storage and work against the production of heat while in use. The ember produced by this method also tends to be small, which leaves it more prone to succumbing to environmental factors. This means it is more likely to be extinguished before it has been transferred into a tinder bundle and encouraged into flames.

Moving away from the equator, the addition of a bow (in order to achieve greater mechanical advantage) appears in historical records – hence the existence of the hand drill's relative, the bow drill. The fire-making methods of Native American tribes illustrate this clearly. Across much of North America the hand drill was used. Head further north into Canada, though, in the days before alternative methods arrived, and you would find many of the tribes employing the bow drill instead. Despite this generalisation, there are some surprising exceptions; the hand drill was even used along the Pacific coast of North America, by both the Tlingit of Alaska in the Sitka area, and by the Bilhula of Salishan stock in the Bella Bella area of British Columbia. Some of the fire sets collected in these areas in the nineteenth century have had the whole of their hearth sticks scorched deliberately in the fire; this treatment would almost certainly have been administered to repel moisture and improve combustibility.

I suspect that the hand-drill technique was mainly for summer use in these areas due to the unsuitable conditions encountered in the winter. It is not an insurmountable task, though, to make fire using this technique in

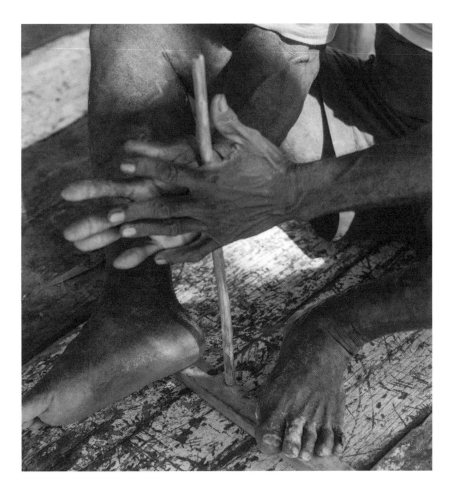

Hand drill in use, West Papua Province, Indonesia.

extreme conditions; on countless occasions I have made fire during the damp, cold conditions of the British winter with relative ease. Two colleagues even made an ember on top of a piece of ice in the high Arctic during winter. Even so, the hand drill's dependability is directly related to the environmental conditions it is used in. If you find yourself ill-prepared and in a survival situation in a cold, damp place, it would be best to opt for a different method. Save your energy for something you are more likely to succeed with – remembering that you're probably not going to be functioning at your best.

While the hand-drill technique remains essentially the same amongst indigenous people around the world, there is a great deal of variety when it comes to the components that make up a hand-drill set.

Across the world, people encountered different materials possessing varying qualities, and this led to variations on the same principle. The Kalahari bushmen in southern Africa often collect their drills from the bush in a dead, dry, straight state and use them immediately. Other people collect their drills from living wood, scraping it down until it is smooth and straightening it in their hands or between their teeth while it dries. These sticks take more time to dry out – although in the African sun, they could potentially be used the same day. The finished result is a superior drill that is a pleasure to use and lasts many seasons.

The hearth sticks have even more variety than the drills. Most of the time a simple, flat board is used, but some use a thin stick with no greater diameter than that of their drill. An unusual hearth set-up was used by the natives of the West Indies, Haiti, the Dominican Republic and Nicaragua. Their approach involved binding two short sticks about the thickness of a man's finger tightly together, laying this down on top of dry tinder and effecting the drilling at the join where the two sticks met. The friction dust either collected on both sides of the drill or fell through the tiny gap and onto tinder placed below.

The notches that are cut into the hearth stick to provide a place for the friction dust to collect are no less variable than the other details already mentioned. Some are deep and wide, while others are very slight, in some cases being little more than a shallow score mark. In Gabon in western Africa, some people didn't cut notches at all; instead they drilled straight into the hearth stick, and the ember formed in the ring of friction dust that collected around the drill. Only certain types of wood will work in this way, which restricts the distribution of this notchless technique. Hearths collected from this region are of a very light wood, resembling the hibiscus employed by the Polynesians.

## HOW TO MAKE A HAND-DRILL SET

Both components of the hand drill have equal importance; one without the other would not achieve fire. The drill, however, needs more attention to detail than the hearth. So long as the hearth stick is made from dead, dry wood in the correct condition and is of adequate dimension, it will produce fire. The drill, on the other hand, needs to be a very carefully selected stick. Tribal people will often dedicate precious time in search of the perfect drill – but they will also be opportunistic. If they spot a potential drill when out and about, they will cut it and store it for another time. While you are learning how to make fire, it helps to get into this habit.

It is important to remember that certain species of wood lend themselves better to making fire using this technique, and different types of wood offer a varying array of difficulty from easy to almost impossible. It would be wise to start with an easy wood before experimenting with others. The most important thing to remember, though, is the wood's condition. When looking for dead wood, the softer varieties that have begun to decompose ever so slightly but are not yet rotten or crumbly are best. You can test the wood's consistency by pressing your thumbnail into it. It should be soft enough to leave a slight mark but not so soft that your nail penetrates into the wood. If you can't make a mark, the wood has only recently died and is still too hard. Some of the best woods for the drill are softer varieties with a core of pith running down their length inside.

**SUITABLE DRILL WOODS**

Alder, aspen, balsam fir, bamboo, baobab, buddleia, burdock, cottonwood, elder (foliage shown ABOVE), false sandpaper raisin, grewia, hibiscus, horse chestnut, ivy, lime, marula, mangetti, mullein, Norway maple, poplar, red flowered kurrajong, reedmace, sage brush, sotol, sycamore, teasel, white pine, willow and yucca.

**SUITABLE HEARTH WOODS**

Alder, aspen, balsam fir, baobab, cedar, clematis, elder, hibiscus, horse chestnut, ivy, juniper, lime, marula, mangetti, Norway maple, poplar, red flowered kurrajong, sotol, sycamore, white pine, willow and yucca.

# HOW TO MAKE THE DRILL

A good, finished drill should be about 1cm thick, 50–70cm long, and straight and smooth for its entire length. I will show you how to make the 'perfect' drill and the 'spliced' drill. The choice depends on what materials are available and how urgently a fire is needed.

## THE 'PERFECT' DRILL

To make a superior drill that will last potentially for years, and is perfectly smooth, straight and blemish-free, you will need to cut a living branch from an elderberry shrub, or if none grows in your part of the world, a similar wood. Unlike dry, dead wood, wet, living wood is supple and easy to straighten – and straightness is important. A drill that is not straight is more difficult to use and will cause blisters.

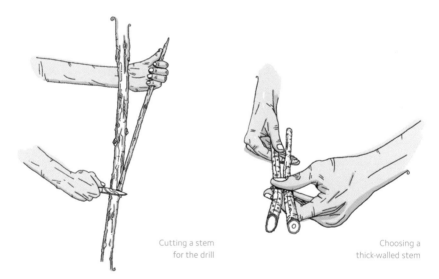

Cutting a stem
for the drill

Choosing a
thick-walled stem

| Look for a stem which is 12mm thick at the base and as straight as possible for at least 60cm. Elder stems have a soft, pithy core which is surrounded by a woody exterior, as do many shrubs. The best drills are those with the minimum amount of pith at their base – the thickest end that will drill. The base of secondary growth as well as the whole of the primary stem has very little pith and is ideal. The nearer to the top of the stem you get, the more the pith quantity increases and as a result, the woody wall gets thinner. To preserve this precious, thick-walled stem, trim it from the shrub as close as possible to the main trunk,

44

right where it emerges. Be careful and take your time to remove it without splitting it. I find a small saw worked halfway from one direction, then finished from the other, works very well. If you only have a knife, make a series of small cuts all around the stem, then snap it cleanly off. Trim off any branches growing from your stem.

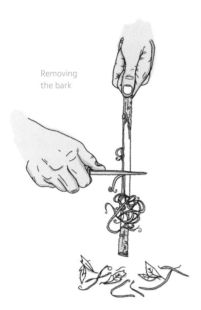

Removing the bark

|| Next, using the back of your knife, scrape along the length of the stem, pushing down hard enough to remove all the bark, until you are left with a clean piece of wood. The surface will be wet at this point, but will soon dry in the air.

||| With a sharp knife, work your way from one end to the other, and carefully slice off any little bumps, giving these areas a light scraping afterwards to ensure smoothness. The last job with the knife is to neaten up the ends. The thinnest end can be trimmed flat, but the drilling end should be a very slightly rounded point.

|||| Now you are ready to start the straightening process. Sometimes, with a bit of luck and some savvy stem collection, it will already be straight and won't need much, if any, work. Most of the time, it will need it. Alternate between looking down the length of the drill as if it is a gun barrel to check the straightness and bending the kinks out of it. Don't rush it. You will get a feel for how far you can push it once you've snapped two or three! When tribal people do this, they use the campfire to help dry the stem out faster, to give the wood fibres some suppleness so it is easier to bend and straighten.

Once you have spent a few minutes on the straightening process and are satisfied, put the stem to one side and come back to it later. Having left it to rest, you will find it is drier. Some of the kinks may have reappeared – if so, bend them out again. This time you will feel a difference, and the drill should stay straight. I have cut stems like this on hot, midsummer days in the

Straightening the drill by bending around the knee

morning, and used them to make a fire the same evening, but usually they are kept somewhere warm and dry for a couple of weeks or so before use. Come back to them every so often and straighten them, and you will soon have the perfect drill.

## THE 'SPLICED' DRILL

Sometimes you may need to procure fire quickly, in which case you should look for a dead, dry stem of correct dimensions. The best solution is to cut a short piece of suitable wood, which is easier to find, and attach it securely to a second piece of any wood in order to increase the length sufficiently to allow the drilling action to take place. The Navaho of New Mexico sometimes used an old arrow shaft for this job. There are several ways of mounting this 'drill bit', but what follows is an approach that can almost always be used.

| Find a piece of suitable fire-making wood, which should be straight for at least 9cm. Carve one end down almost to a point, making the taper at least 7cm long and ensuring it has a perfectly square cross-section with four clean sides. Put this drill bit to one side.

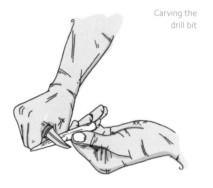

Carving the
drill bit

|| Select any stem from the forest, the only important factors being that it should be straight, of the correct diameter and long enough to make a drill.

||| You need to prepare this piece to accept the drill bit. Start splitting the thickest end in half with your knife, but do not let the split run any further than 10cm. Next, carefully make a similar split, but this time at 90 degrees to the first, so as to quarter the stem.

|||| Having split the end of the stem into four pieces, you will have a 'socket' which will receive the tapered drill bit nicely.

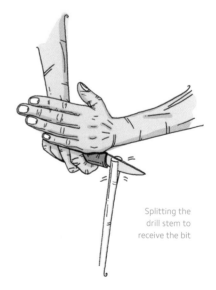

Splitting the
drill stem to
receive the bit

46

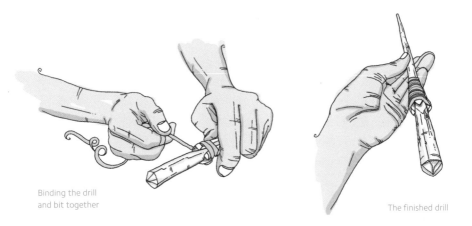

Binding the drill
and bit together

The finished drill

||||    Having inserted the drill bit, let it sit naturally in the socket, with the corners of the bit resting in between the split ends.

|||| |    To finish the drill and make it ready for use, it must be bound tightly. Use a long, flat strip of willow bark about 6mm wide (or something similar).

|||| ||    Once bound tightly, finish it off by tapping the bit a few times to force it further into the socket.

## HOW TO MAKE THE HEARTH

It is surprising what can be used as a hearth. Often the most unlikely looking stick can be employed if it is dry and handled skilfully. While learning, it is best to select a stick that can be carved into a flat, wide board, ensuring all sides are as square as a plank. It should be as thick as the diameter of the drill (1cm), at least 3cm wide and at least 20cm long. You need to be able to hold the hearth stick steady as you drill. This is usually achieved by pinning it to the ground with your foot. The vigorous drilling action is apt to wobble the hearth, and this will scatter the friction dust. Carve it perfectly flat, so it sits flush with the ground and remains steady, allowing the dust to collect nicely.

Hearth made from willow wood.

Making fire with the hand drill is very easy; the challenging part is the learning process. It is not the impossible task that some will have you believe, but you do need a degree of fitness and good technique. Do not be disheartened by a lack of success. It is necessary to go through this in order to learn.

There are many ways to prepare the components of this fire set, but the following method is the most reliable when you are learning.

| Using the very tip of your knife, gouge a slight depression a little greater than the diameter of the drill tip into the centre of the hearth board, close to one end. The purpose of this is to give the drill a seat to prevent it from slipping out when you first begin to drill.

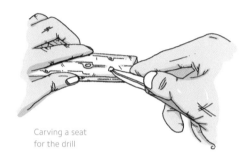

Carving a seat
for the drill

|| Find a comfortable position that still allows you to exert downward pressure onto the hearth with the drill. Place the drill tip in the depression.

||| Now the key to the technique: moisten your hands! Use either spittle or water. Rub your hands together a little until you feel them become 'grippy'.

|||| Starting at the top, begin to spin the drill between your palms while applying a simultaneous downward pressure. Don't let the drill roll onto your fingers. As you do this, your hands will inevitably move down the drill. Keep going until they reach the bottom. With one hand holding the drill securely in the depression, the other hand can swiftly move back up to

Seating the drill
into the hearth

the top, the thumb hooking over the top of the drill to apply pressure, allowing the other hand to slide up to the top too. This process is repeated swiftly over and over.

卌　After two or three of these passes, smoke should pour from the point of friction. Do a couple more, then stop, but don't exert yourself too much. You are merely seating the drill in at this stage.

The notched hearth with
a wood shaving placed beneath

卌 |　Now, cut a one-eighth segment from the side of the board into the middle of the scorched hole you have made. Imagine you are cutting a perfect one-eighth slice of cake; the point of the notch should be directly in the centre of the hole.

卌 ||　Now you are ready to make fire. Place a thin shaving of wood or a dry leaf under the notch, and clamp the hearth to the ground again. The friction dust will collect on this shaving instead of the ground. This gives the developing ember some protection from possible dampness and also provides a means of transporting it to your prepared tinder bundle once the time comes.

卌 |||　When everything is ready, spit into your hands again and start drilling as before. Start in a relaxed but determined manner, gradually increasing speed and downward pressure as you see the quantity of smoke increase.

After three or four passes down the drill, the smoke should encase the bottom of the drill in a thick swirl. When this happens, watch the notch as it fills up with friction dust and oozes its excess out onto the shaving. When you see smoke coming from underneath the little pile of dust, stop drilling.

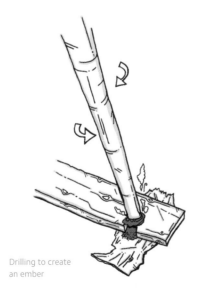

Drilling to create
an ember

### IGNITING THE TINDER

Having made an ember, you can now use it to make a flame.

I   Carefully roll the hearth board away from the wood shaving supporting the smoking pile of dust containing your precious ember. This is a delicate process. You might need to use a small twig to hold the dust in place as you roll the hearth away.

II   Transfer the ember to the middle of your tinder bundle – ideally where it is driest and finest – and wrap it up gently.

III   A few gentle breaths should be enough to intensify the heat of the ember and ignite the tinder into a burst of flame.

# CHAPTER 3

# BOW DRILL

**OVER THE PAST TEN YEARS I HAVE TAUGHT HUNDREDS OF PEOPLE** how to make fire by friction, mainly using the bow drill method: people of all ages, from children aged seven years up to those in their late seventies. It doesn't matter what your background is; if you are new to making fire, you are on a level playing field. Everyone encounters the same problems, setbacks and challenges, and no one has an unfair advantage when it comes to the tricky job of building a fire and surviving in the wilderness.

I once taught Jake, a former Army soldier who had fought and been severely injured during an ambush while on duty in Afghanistan five years before, how to use the bow drill. Jake had attended our basic course the year before, during which he had been introduced to the bow drill. This was a more advanced course.

At the beginning of the session, I recapped the principles of the bow drill, giving everyone the collective task of making a bow-drill set together and, ultimately, a fire. I immediately warmed to Jake's positive attitude, easy-going smile and great humour. He was extraordinarily keen to get the set up and running because, as I discovered, he had not had much success at making a fire.

'Dan, if I leave this week with no other skill than being able to make fire by friction, I'll be happy,' he grinned, as he got his kit together. Jake was so enthusiastic that he had made a set for himself, as well as helping make the group set. As soon as he had finished carving, he knelt down and bowed vigorously, but the drill kept flicking out from the hearth. Time and time again, he patiently replaced the drill into his bow and re-started the process. Finally, the drill flicked out with even more force and came to rest in some nearby brambles. Jake stood up and threw his bow hard into the undergrowth, cursing. His hands shook with fatigue and frustration.

'My arm doesn't work any more,' he barked, taking a long draw on his roll-up. He threw the butt into the campfire, stood, and went to retrieve the bow and drill from amongst the leaves. By the time he returned, that familiar glint had entered his eye.

'Right, I'm gonna do it this time.' Chuckling to himself, he knelt down again and started to bow.

I could see that he needed to impart more downward pressure onto the drill to increase the friction. This is something that most people need to develop when they learn the technique. Jake was just as close as anyone else, but his injuries meant that he was at a real disadvantage.

'Jake, you're not getting enough pressure downwards. I have something I want to try, if you're up for it?' I said. I remembered there was a simple way round the problem that could be employed if someone had injured their arm in a survival situation. I asked him to go out and cut an 8cm-thick sapling and keep it long – at least 2m. When he brought it back, I cut a shallow depression in one end to receive the top of the drill. We made a giant bearing block which not only added a bit of extra weight on top but also offered more stability; it allowed Jake to rest all of his left forearm on the branch instead of holding a small block precariously in his hand.

After only a few moments using this new set-up, he had an ember. Shortly after, a blaze was roaring on the ground and Jake stood beside it with a broad grin, smoking a celebratory roll-up. This feeling resonated with me; sometimes it's the smallest accomplishments that can feel like the monumental victories.

## USE OF THE BOW DRILL

The term 'bow drill' is slightly restrictive – perhaps a more accurate description for the technique would be the 'cord and drill'. However, bow drill is now the most well-known form. The cord used to achieve the spinning action of the drill was not always in the form of a bow; in fact, more often than not, a length of cord fitted with a toggle or handle at each end was employed. A relative of the simpler hand-drill method, the addition of a string to achieve greater mechanical advantage occurs most prominently in the extremely cold northern regions of the planet – places where fire was often more difficult to make, and the ability to do so was often the difference between life and death.

While it is no longer used by any tribal group today, we know it was certainly not restricted to the colder areas. Indeed, in the past it had an impressive range, being recorded in ancient Egypt, Mesopotamia, Sri Lanka, the state of Perak in the Malay peninsula, and amongst the indigenous people of the Mentawai Islands off the west coast of Sumatra. In the high latitudes of the north, it was heavily relied upon by the Inuit, and there are accounts of its use further south amongst the Sioux and other

North American tribes. I suspect, however, that the technique had an even wider range than we can prove and that in some areas, particularly the boreal forest with its sultry summers and icy winters, people switched between this technique and the hand drill according to the season or the weather at the time. In the Arctic and sub-Arctic north, we know that it was employed from mainland eastern Russia and amongst the scattering of islands in the Bering Sea (including the Aleutian chain) eastwards, across most of Alaska and northern Canada, and further east into Greenland.

Today, the bow drill is considered the most versatile method of making fire by friction. Most species of wood will work, because the cord provides a mechanical advantage. This offers the ability, when needed, to drill hard and apply a prolonged, controlled effort – something that can be more difficult to achieve with other methods. Secondly, in a survival situation when exhaustion, dehydration and injury are more likely to be encountered, a pair or even a small group of people can team up and share the workload – again, something that is not always possible with other techniques. Even group members who have no prior knowledge of the technique can assist. Finally, the use of a cord allows you to spin a thicker drill easily. This results in the production of a larger ember, which is in turn better able to withstand unsuitable atmospheric conditions and substandard tinder. For these reasons, in a crisis or survival situation, this method gives the best chance of producing fire wherever on earth we may find ourselves, providing there is wood to make it from. If you're going to learn only one method, make sure this is it.

Inuit using a bow drill.

## HOW TO MAKE A CORD-AND-DRILL SET

Essentially, the cord-and-drill method is a slightly modified hand-drill set with the addition of a bearing block and a cord. There are four parts to the set: the drill, the hearth, the bearing block and the cord – either fitted with toggles or securely fastened into a bow. The only parts that need to be given careful consideration in terms of material selection are the drill and hearth, because they are the two pieces that work together to make fire.

### SUITABLE WOODS

The majority of woods will work but try to find one of the following before resorting to others: alder, ash, aspen, baobab, cedar, elder, hazel, hibiscus, horse chestnut, ivy, lime, mangetti, poplar, red flowered kurrajong, sotol, star chestnut, sycamore or willow.

If one of these varieties cannot be found or if you are unfamiliar with the local species, look for the usual dead, dry, standing wood, carve its surface a little and as a general rule choose a softer, easily-carved wood over a hard variety. Species that have a pithy central core also tend to work well, but remember whatever you select must be strong enough to withstand the stresses involved in the drilling action.

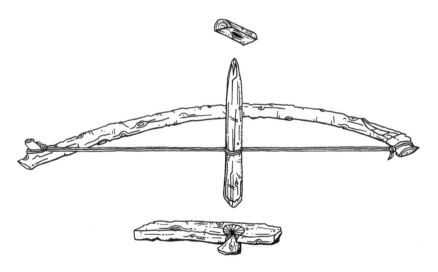

**FROM TOP** Bearing block, drill, bow, hearth board.

As with other techniques, there are some slight differences in the dimensions of the individual components from one place to the next according to the environment, the maker's tools and the materials available. All things considered, the following guide describes the dimensions for the most versatile set that will work anywhere.

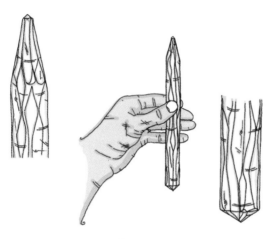

THE DRILL

The drill for this set is shorter and thicker than that of the hand drill, the best size being approximately 25cm long and precisely 2·5cm thick. As the downward pressure will be applied with the aid of a wooden bearing block, the top of the drill needs to be shaped in such a way as to minimise friction at that point. Carve the top of the drill into a tapering point, like sharpening a pencil. The very tip of this point should not be sharp but slightly bevelled for improved durability.

The bottom of the drill is the business end, and will be in contact with the hearth, so it should be shaped in such a way as to maximise the friction. Carve this end of the drill to a flattish point: a slight point is necessary to help the drill bite into the hearth and prevent it slipping out while in use. It is essential that the drill is perfectly straight, so really take your time to choose the straightest section of wood from the branch. Then you need only shave off a few layers of wood to smooth the surface, instead of struggling to carve a straight drill out of a bent piece of wood. The drill should have parallel sides apart from at the ends.

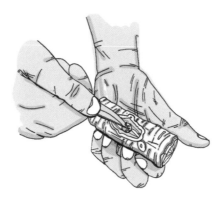

THE BEARING BLOCK

You need to be able to impart considerable downward pressure on the top of the drill. You therefore need a block of wood that can be held in one hand, into which the top of the drill sits. The aim is to minimise the friction as well as slow the boring action of the drill against the block. The top of the drill is already suitably carved, but you can further improve function by choosing a hard piece of wood that is still living. The higher moisture content will help to prevent burning and the hardness will slow the erosion of the block. It is by no means essential to choose a piece of wood with these qualities but it does make the process easier.

There could be no better example of this than the finely embellished bearing blocks used by the Inuit; they were beautifully made from wood, ivory, the vertebrae of large fish or the talus bones from caribou. Some even had a smaller block carved into the top to act as a mouth grip – the block being held in the teeth of the user in order to free up both hands, and the hearth resting in the lap. Quite often a small piece of soapstone, obsidian or marble was shaped into a socket and polished before being inlaid into the block. Acting like a jewel in a fine mechanical watch, it reduced abrasion and

Freshly carved, unused bearing block.

extended the lifespan of the set. Of course, in an emergency it is not necessary to go to such lengths; all that is needed is a suitable piece of wood cut
slightly longer than the width of your hand, smoothed of any branches or
rough areas to ensure comfort, and a 1cm-deep depression carved into the
underside. This needs to be big enough to accept the top of the drill and
prevent it from slipping out while drilling.

### THE HEARTH

The hearth boards made by tribal peoples varied in design. One of the most
interesting was an Inuit 'stepped' version, meaning a step was carved onto
the side of it to act as a barrier to prevent the ember falling onto the snow.
Other approaches avoided cutting the usual vertical side notch; instead,
people drilled into the middle of the hearth directly onto a deep lengthways
groove, with the ember forming in the groove instead. Some hearths show
a line of overlapping scorched holes – the ember forming in the previous
hole. These clever techniques had evolved from the much simpler hearths of
warmer climates as a result of the extreme environments, and are well worth
considering if you are ever forced to make fire by friction in these areas.

    The hearth will need to rest on the ground without wobbling
or rattling, so you will need to carve it flat. The finished board should
be rectangular in cross-section, with all sides as square and as neat as
possible. Make it precisely 2cm thick, and wide enough to accept the
diameter of the drill with plenty of margin for error. 4cm is a good guide
for width. In terms of length, there needs to be at least enough room

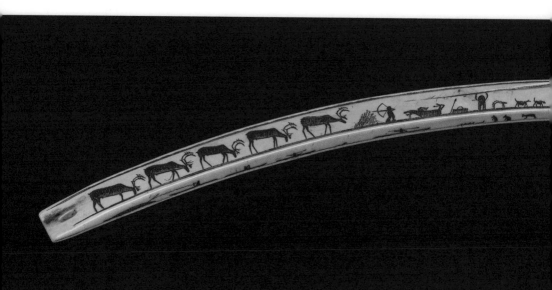

to accommodate the diameter of the drill plus the width of your boot. Ideally, though, you should make it longer than that, to ensure there is space left over for several attempts at making fire – 30cm is good. As you carve the hearth, check that it is flat for all of its length and has no twists in it. Imagine you are carving a miniature floorboard.

## THE BOW

If you are making fire on your own either during training or in a genuine emergency, you'll need to make a bow, as it is far easier than using toggles. The bow should not flex at all and should be as stiff as possible. Again, the perfect examples are the bows that were once employed by the Inuit – these were commonly made from a walrus tusk that was carved down or the fibula of a caribou. Wooden bows were also used, and this is the best option in most areas.

I    Cut a branch that is 2cm thick and 75cm long with a slight curve. It can be dead or living wood as long as it is not brittle, too heavy or flexible. The way the bow is prepared depends on what the cord you intend to use is made from. Nylon cord and similar materials can withstand knots being tied in them while under strain; natural fibres, on the other hand, need to be treated with more care, so we must secure them to the bow in a different way.

II    If you are using nylon cord, to prevent it from slipping down the bow you'll need to carve little notches on the outside of the curve 2cm in from the ends. If you can find a branch with a natural fork at the thinnest end, use that to save you carving one of the notches.

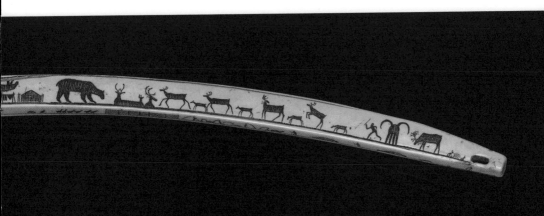

Painted ivory Iñupiaq bow, c. 1880–1920, Kotzebue Sound, Alaska.

It doesn't matter what knots you make but start by making a fixed loop in one end of the cord and slip it over the thin end of the bow so it sits in the fork or carved notch. On the other end of the bow, it is best to make a secure knot that can be easily untied when needed – inevitably the cord's tension will need adjusting throughout the process. A few round turns and two half hitches is all that is necessary. Keep the string slightly slack; if it is tight, you will not be able to twist the drill into it.

||| If you are using plant or tree fibre cord, using some of the leftover fibres gathered for the production of the cord, make a secure binding 2cm in from each end of the bow. Then, using your knife, split the ends of the bow up to the binding. You can now push the ends of the twisted and knotted cord into these splits. Once the drill is twisted into the bow, it will pull the cord tight and the overhand knots will prevent it slipping through the splits. If you need to tighten the cord slightly, remove one end and give the entire cord a couple of twists to shorten it. If it needs more than a slight adjustment, consider re-making the overhand knot.

## THE CORD

In the past, a variety of materials were used as the cord to spin the drill. The most resilient of the traditional materials is rawhide, as employed by the Inuit. Far north of the tree line in the high Arctic, there was little else that could withstand the vigorous action while maintaining flexibility. Further south, however, and particularly in the tropics, plant fibres were readily available, and although they tended to have a shorter life span, they were easily replaced. With a wide variety of strong plant fibres to choose from, people made thongs of rattan fitted with toggles or twisted tree bark into lengths and strung them in bows.

It is best to use a strong nylon cord about 4mm thick, especially when first learning this technique. As always, though, it is important to know how to do without it. If you do find yourself in such a situation, you can use a bootlace, or cut two or three adequate lengths in a spiral from the bottom of your trouser leg before rolling these together into one thicker cord. If that is not possible, look for plants or trees that provide suitable fibres. Willow bark is very good for this task, as is wych elm and lime. There is usually a tree that provides bark in a similar way to these species whatever environment you are in, so it pays to do some research before you venture out. Prepare the wood as follows in order to make your cord.

|    Try to select a branch that is at least 3cm thick, as straight as possible for 1m and with as few side branches or knots as can be found.

||    Scrape off the outer bark if necessary and remove the strong inner bark. You will need three 1m lengths, each being about 8mm wide.

|||    Lay the lengths alongside one another so they are stacked in a flat pile and tie an overhand knot at both ends.

||||    Now, treating the three lengths as one, roll them in one direction to encourage them to form a round cord and don't let them unwind.

||||    Another option is to strip the fibres from three or four stinging nettles and treat them in the same manner. If you cannot find a tree with suitable bark, or if it is difficult to remove it from the branch, look just under the ground's surface for thin, flexible rootlets. Find one that is about 6mm thick and as straight as possible for 1m. Remove the outer rind from it. Then, securing one end under your boot, begin to twist it all the way down its length as you would if you were making a withy (thin, wispy tree branches twisted along their length to make them pliable and usable like wire). If you twist a rootlet in this manner, it increases its strength and flexibility quite considerably, and works very well as a cord to spin the drill.

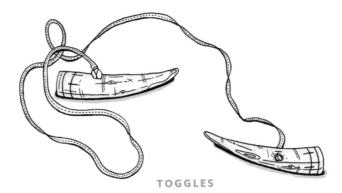

TOGGLES

If there are two or more people in your party, toggles are a pleasure to use and in some ways easier for a novice than a bow. The Inuit made great use of this method; their toggles were sometimes simply made from plain pieces of wood, while others were beautifully crafted from large sea lion teeth, complete with detailed embellishment.

To make toggles today, all that is required are two small sticks about 1·5cm in diameter and 10cm long. They should be strong enough and smoothed to ensure comfort when in use. Attach them to your cord, one at each end. It is worth stating that, while easier to use, toggles are not essential. The Inuit sometimes used a strip of seal skin on its own without any toggles at all. Fire can be made by holding the ends of the cord. This is particularly useful to know when it comes to using plant and tree fibres, as there is no need to tie any knots.

## HOW TO MAKE FIRE USING A BOW

Having prepared all the components, the set needs to be 'seated in' before it can be used to make fire. This process only takes a minute.

I    Using the very tip of your knife, gouge a slight depression a little greater than the diameter of the drill tip into the centre of the hearth board, close to one end. The purpose of this is to give the drill a seat to prevent it from slipping out when you first begin to drill.

II    Place the hearth on a flat, level piece of ground away from snags, and secure it to the ground by placing the ball of one foot on top of it, close to the depression.

III    Slide your other leg backwards and place your knee on the ground where it feels both stable and comfortable. Bear in mind most of your body weight should be over the foot that holds the hearth in place.

||||     Turn the drill into the cord so that it lies on the outside of the bow. This will allow the most efficient use of the bow.

||||     Put a pinch of green leafy material into the depression in the bearing block. This will help to lubricate the workings and reduce friction.

|||| |     Place the bottom of the drill into the depression in the hearth and place the bearing block on top. The hand that holds the block needs to be kept absolutely stock still against your shin, so adjust your body position accordingly.

|||| ||     Hold the bow right at the back end and curl your fingers around the tight cord. If the drill starts to slip during the process, a temporary measure is to squeeze the cord against the bow with your fingers in order to tighten it. It is also more stable to hold it in this way.

|||| |||     For tandem use, have your partner facing you on the other side of the hearth and mirror your position. They can place one hand on top of yours on the bearing block for extra stability, while their other hand can hold the other end of the bow.

|||| ||||     Now you can begin to drill by swinging your arm from the shoulder. Move the bow steadily backwards and forwards using its entire length. Ensure the drill is kept vertical and the bow kept at 90 degrees to it at all times. Keep drilling until you have produced thick smoke for about twenty seconds. Carefully lift the drill and inspect the hearth. It is ready when a circle of the same diameter as the drill has been scorched into its surface.

|||| ||||     As for the hand drill, cut a one-eighth segment from the side of the board into the middle of the scorched hole. The point of the notch should be directly in the centre of the hole.

||||| ||||| | Place a thin shaving of wood or a dry leaf under the notch and clamp the hearth to the ground again. The friction dust will collect on this shaving instead of the ground. This gives the developing ember some protection from possible dampness and also provides a means of transporting it to your prepared tinder bundle once the time comes.

||||| ||||| || When everything is ready, twist the drill into the bow again and start drilling as before. Start in a relaxed but determined manner, gradually increasing speed and downward pressure as you see the quantity of smoke increase. After half a dozen swings of the bow, smoke should encase the bottom of the drill in a thick swirl. When this happens, watch the notch as you drill, and you'll see it fill up with friction dust and ooze its excess out onto the shaving. When you see smoke coming from underneath the pile of dust, stop drilling. This usually takes around thirty seconds but will vary depending on the type of wood you are using.

||||| ||||| ||| From this point onwards the ember you have made should be handled as usual (see page 50).

## HOW TO MAKE FIRE USING TOGGLES

The process of making fire with toggles and cord is much the same as when using a bow, the one difference being that each person is dedicated solely to their respective task. One person will hold the drill steady and impart downward pressure with both hands on the bearing block, while the other holds a toggle in each hand, and rapidly moves their hands backwards and forwards to spin the drill. You'll notice that the string will slip more easily when using this technique, and it takes a bit more getting used to. It helps if you make two turns around the drill with the cord instead of just one. Also, be sure to keep the cord taut at all times; if it is allowed to go slack it will slip.

As always with these techniques, until you get the hang of it there will be frustrations and failure. If you persevere, though, it will become easy and you'll realise it is a fair exchange; you will be able to make fire anywhere with what nature provides.

OPPOSITE Fire from a bow drill set made with British wood.

CHAPTER 4

# FIRE PLOUGH

**AS WE SAT IN A SMALL CLEARING IN THE BAINING MOUNTAINS OF** East New Britain Province, Papua New Guinea, I knew we were about to witness something unique. The Milky Way shone magnificently above us; aside from a few torches held by boys from the village and a central blazing fire, the area was completely dark. I sat enthralled, waiting for one of the local fire dances to start.

It was 2017, and I was on my latest quest to discover more about how mankind makes and interacts with fire. The Baining people are among the original and earliest inhabitants of the Gazelle Peninsula of East New Britain, and are thought to have been driven into the mountains, from which they get their name, by the nearby Tolai people and by volcanic activity. As recently as 1994, the provincial capital of the area at that time, Rabaul, was completely destroyed by simultaneous volcanic eruptions from the active volcano Tavurvur and its geological twin Vulcan, which buried the town under several metres of ash. I had been keen to visit this area to see both their use of the fire plough to make fire and their legendary fire dances. Seeing photographs and reading tantalising tales from a land that is still shrouded in mystery and rarely visited was enough to make me start the planning process.

The Baining people traditionally conduct their dramatic dances to mark special occasions: to celebrate the birth of a new child, to mark the commencement of their harvests, to remember their dead, or as a rite of passage when initiating young men into adulthood. Sometimes, large feasts are prepared where taro, pigs, pythons and cassowaries are cooked and eaten before the dance. Anyone can spectate, but only initiated men can observe the 'secret place' in the bush where the dancers adorn themselves in preparation for dancing. They wear special masks, and women and children must not see them apart from when a dance is taking place.

During my trip to New Britain it had rained heavily every afternoon, and continued to pour well into each night. As is often the case in the tropics, the rain was reliably regular, and I was worried that the nighttime fire dance would be disrupted or even cancelled. The day of the dance was no different; the rain started in the early afternoon and kept everyone under cover. The villagers took the opportunity to cook us some vegetables they had grown in their fertile and beautifully kept gardens. Stones were heated on the fire before being used to cover packages of food wrapped carefully in banana leaves. A couple of hours later, while we were eating, the clouds unexpectedly parted and we enjoyed relaxing amongst the last rays of the setting sun. Darkness cloaked the land swiftly as it always does this close to the equator, and we were treated to the most exquisite night sky I have ever seen. My Baining guide David explained that a special spell had been cast by a spirit man during a secret ritual in order to influence the weather. He didn't elaborate nor, out of respect, did I pursue any further explanation.

David appeared from the shadows clasping a small smouldering log in his hand which he placed amongst some dry palm leaves in the centre of the clearing. A few local boys squatted around it with me. When it suddenly burst into flame it illuminated their faces, and I could see the anticipation in their eyes. The boys began to lean piles of wood up next to the fire, and as the flames grew and started to lick the night sky, a group of a dozen or so men, each carrying a long tube of bamboo, and who up until then had been mingling in the shadows, started to arrange themselves into three or four rows on the periphery of the clearing. Suddenly, they began to thump the ground with their simple bamboo instruments. The complex rhythm complemented their chorus of loud and uplifting chants.

Then a murky, shamanic figure appeared on the edge of the clearing, assessing the scene before leading the dancers out one by one into the flickering light cast by the fire. The men ran straight out, and right up to me where I sat, next to the choir. One wore an elaborate mask made from tree-bark cloth, stretched taut over a bamboo frame, and painted eerily with large circular eyes and traditional patterns using pigments from wild plants. Leaves draped down from his neck and shoulders like an oversized ruff. Two other dancers wore pointed, orange hats, their faces hidden by long, narrow leaves hanging down from the rim. Long, wispy canes decorated with white feathers stuck out of the top of each hat. They were all bare-footed and naked apart from leaves bound tightly to their upper arms and calves, encasing the latter completely. Their upper legs and forearms were smeared with white paint. They looked like creatures from another world. Their outfits are said to be representative of the spirits dwelling in the forest. David explained that the design of their unmistakeable mask originated from a dream experienced by one of his ancestors long, long ago. Traditionally, their costumes are used only once for the ceremony before being thrown into the fire.

The masked dancer held the head end of a python and a young boy followed on behind holding the tail, rather like a bridesmaid holding a train. Swinging the snake violently, they moved to the beat of the bamboo and the resonating chants. It seemed to rouse something inside the dancers, and they moved with extraordinary energy as if they were possessed.

The heat from the fire was searing, and even though I was several metres away from it, sweat poured off me like water. The scene playing out before me had been repeated on countless occasions over thousands of years – a powerful concept. The anticipation of what was to follow seemed to affect everyone who was present; the electricity in the air was palpable.

As the spectacle continued and the music reached a frantic pitch, the masked dancer careered into the fire, releasing a million sparks up into the air, with nothing to protect him except the magic of the dance. Up and up went the sparks, rising high on a powerful thermal column pushing up into the darkness. Mingling as they ascended, it was as if a huge swarm of fireflies had been set free. The brave dancer emerged apparently unscathed on the other side of the fire, scattering dozens of glowing coals across the ground, and stamping on the glowing embers with his bare feet as they settled. The flames seemed to suffer a little, but soon regained their vigour, and before I knew it, the dancer was back, committing himself to repeating the act, this time lingering in the centre of the fire for a few seconds, the flames devouring his bare legs, before exiting and continuing to dance as before.

The evening drew on. It was as if I was transported to a different time and place. Eventually, the huge stack of firewood the men had prepared burned out, and the embers were scattered across the clearing. Suddenly all was dark, but some of the bare-footed young children ran around wildly, continuing to stomp on the embers and kick them around as if they were training for the future, like young bucks testing out their first set of antlers.

## THE FIRE PLOUGH

The fire plough is still found today in parts of the jungles of central Africa, particularly in the Kasai region of the Democratic Republic of the Congo, and further north along the banks of the Congo river. This is a stone's throw westward from where evidence suggests early hominids first appeared. Although through most of the rest of Africa the hand drill was used, in the Congo the plough has persisted. Some say the reason for this is the termite's preference for attacking woods that are more suitable for the hand drill. Apart from its employment in the Congo region (as well as by the Ngarla people in the desert of western Australia, around present-day Marble Bar and Telfer), the plough is predominantly a technique of the Pacific islands. It starts to be found around the Bismarck Archipelago just off the eastern coast of the great island of New Guinea. There seems to be a dividing line in this area; on the mainland in what is today Papua New Guinea, the fire thong is predominantly used, but travel across less than 100km of ocean to New Britain, and the thong is replaced by the plough. South-east from here, throughout the Solomon Islands, Vanuatu, New Caledonia and all the way to New Zealand, the fire plough dominates. It is also found among countless islands scattered across the vastness of the Pacific Ocean to the east and north-east of New Guinea: in Samoa, Fiji, Tahiti and Hawaii.

Modern use of the fire plough across these territories varies. In remote areas of New Britain and New Ireland, and I dare say elsewhere, it is still heavily relied upon. In other places, it may be called upon on the odd occasion when matches and lighters run out. In some areas, it has become redundant and unnecessary from a practical point of view, but continues to be practised and passed on to younger generations as an important part of cultural identity, or for the provision of exciting demonstrations for visitors. During my time in the Bismarck Archipelago, I witnessed people using the fire plough in many villages. The people of New Ireland in particular genuinely rely upon it, even close to the capital of Kavieng. They keep their fire sets tucked up in the rafters of their kitchen huts, ready for action. Even the teenagers, boys as well as girls, were very adept and it was an everyday part of life for most.

In East New Britain, things were different. The young men I met knew about the skill, but it seemed they had never had to use it. In fact, on my trip to see the fire dance I provoked quite a spectacle when I asked if someone could demonstrate the plough for me. The young men put much effort into showing me but were without success, until an older man – the spirit man who had stopped the rain – came along and showed us all. He made fire easily in only a few seconds. The whole community went mad, and began rubbing any old pieces of wood they could find around them together, in an effort to emulate what they had seen. Men, women, young and old, were breathing life back into the old way.

## HOW TO MAKE A FIRE PLOUGH

As always when constructing your fire-making equipment, look for dead, dry, standing wood of the correct hardness. Across the fire plough's range, there is one wood that is used more than anything else: coastal hibiscus (foliage shown ABOVE). This wood is so well suited to the technique that it is possible to make fire in a few seconds. One early account from the Hawaiian islands even states that the resulting ember sometimes actually bursts into flame if there is sufficient breeze – but this is not a usual occurrence. It does, however, demonstrate the suitability of hibiscus over other woods. It is extremely difficult to make fire with anything other than hibiscus or a small handful of others, listed below. If you cannot find the woods in the list, you will need to experiment for yourself. Although hard/soft combinations were and still are used in some places, using the same wood for both components is most common. In the Bismarck Archipelago, it is worth noting that some people will make a fire plough set entirely from bamboo. This is a last resort, however, so is reserved for use if there has been prolonged rainfall and other wood is likely to be wet. Most of the time these people use bakes, patma, ramosaqa or sell which are very soft – almost as light and soft as balsa wood.

Be sure, therefore, to collect suitable wood if you are going to learn this method, or to rely upon it in a crisis. This reinforces my point that unless you are already skilful in using the fire plough and find yourself cast away in a place with plenty of suitable wood, the cord-and-drill method is the best technique to resort to in a survival situation. The benefit of the plough is that the wooden components need very little preparation, and you can make fire from scratch in less than five minutes.

### SUITABLE WOODS

Alder, bakes, bamboo, hibiscus, ivy, kaikomako, lime, mahoe, patma, poplar, ramosaqa, sell, sotol, sycamore and white willow.

### THE PLOUGH

The finished plough should be straight, and 25cm long, 2·5cm wide and 2cm thick (tapering to 1cm thick at the tip). Start with a straight, knot-free piece of wood and carve it down to the dimensions stated. It should taper down to the working end of the plough, which needs to be carved to a 60-degree point.

Tip detail:
side and top view

## THE HEARTH

The hearth needs very little preparation, and providing it is a suitable species, dry and of the correct hardness, any piece of wood can be used. As long as it is large enough to accept the ploughing action and can be held still, anything from a fallen tree to a small stick can be employed.

| If you are using a branch that is not straight, put it on the ground and find out which way it sits steadiest.

|| Using a knife, carve a flat area at least 15cm long and 3cm wide on the uppermost surface of the stick.

||| Leave one of the resulting curls of wood attached at the furthest end of this short, flat area to act as a backstop. The friction dust will collect against the base of this shaving later.

|||| It is not necessary to cut a groove to guide the plough but it can help when you are learning. If you do, make it very shallow.

**ABOVE** Carving the hearth.
**OPPOSITE** Fire plough set in hibiscus wood, New Ireland Province, Papua New Guinea.
Note the hearth was originally longer but has been sawn short for the author's collection.

# HOW TO MAKE FIRE USING THE FIRE PLOUGH

| Ensure the hearth stick is secure – it will not work if it wobbles, twists or moves. The best way to achieve this is to make the hearth stick as long as possible, and kneel at one end with the stick between your legs for solo use, or with the assistance of someone else kneeling at the other end if you are a pair.

|| Clasp the plough securely in both hands, with the thumbs crossed underneath the stick and the fingers of one hand overlapping those of the other, above the stick. Place the tip of the plough on the flat surface of the hearth, close to the base of the curled shaving, and without bending your arms too much, begin to plough backwards and forwards very slowly, holding the plough at an angle of about 30 degrees. As you do this, sway your body slightly from side to side so as to dig the corners of the end of the plough into the hearth. This will form two tram lines, which will help guide the plough and prevent it slipping off the edge of the hearth. The ploughing stroke should be no more than 10cm in length. Take your time and go slow; the aim here is to seat in the plough.

||| The ploughing action should not rapidly tire your arms. The effort should come from your back, and you should bend at your waist – your hands and arms are merely utensils to grip the plough and transfer the force to it.

|||| Apply most of the downward pressure into the forward strokes as you gradually begin to increase your speed and effort. Continue to sway your body from side to side as you plough. You should start to see a groove appear, which will begin to scorch and produce copious amounts of smoke. At this point, you will be producing friction dust with every stroke.

卌 If all you have created is a shiny, polished effect on the wood, it is a sign of insufficient downward pressure. You will not get any further unless this effect is removed. Either add a pinch of sand onto the hearth before continuing, or plough extremely hard in a short five-second burst. Once the surface has begun to char, it is unlikely to polish again.

卌 | The real skill with this technique is being able to plough vigorously while maintaining absolute control. In order for an ember to form, the friction dust must collect in one

OPPOSITE Fire plough in use, New Ireland Province, Papua New Guinea.

place without being knocked and disturbed by the tip of the plough. If you are a pair, one person can nurture the friction dust with their fingertips as the other produces it. This is easiest if you have experience with the other friction techniques, as you will know what is required. If you are on your own, you'll need to take extra care.

||||| ||   Once you have created an adequate pile of friction dust, you must change tactic. Without ceasing the ploughing action, bend your arms, lean forward and this time plough without swaying from side to side. Plough as fast as you can, with a shorter stroke of around 5cm and with as much downward pressure as you can, for at least five seconds. You should see an increase in smoke and a darkening of the dust being produced. Do not knock the pile of dust apart! Until you gain some experience with this method, it is difficult to identify the right time to stop. Stop either if you see the pile of dust smoking on its own or when you are exhausted.

||||| |||   From this point onwards, the ember you have made should be handled as usual. Fan it with your hand, and wait for it to glow before transferring it to tinder and blowing it to flame. Across most of the fire plough's traditional range, an old, dry coconut husk is teased apart and used for this job, but of course similar tinder can be employed in other areas. It is worth noting that sometimes the ember is very small and adheres to the hearth stick tenaciously. In these cases, it is sometimes better to leave the ember where it is and place some tinder in contact with it, blowing it to flame.

When mastered, the whole technique should take no more than forty seconds to produce an ember. Think of it in two parts, each requiring a different tactic. Put simply, the first part – around 80 per cent – is an enthusiastic jog, during which you build up friction dust without tiring yourself out too much. The second part – 20 per cent – is an all-out sprint to the finish, with all of your energy directed into just a few seconds of rapid activity in order to ignite the dust and form an ember.

OPPOSITE Baining people making fire with the fire plough, East New Britain Province, Papua New Guinea.

CHAPTER 5

# FIRE SAW

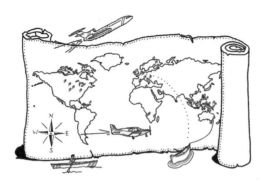

**MY PASSION FOR EXPLORING THE NATURAL WORLD HAS TAKEN ME TO**
the wildest corners of our planet. For me, the best type of holiday is deep
in the wilderness and far from the noise and chaos of civilisation; only
then can I truly unwind and recharge my batteries. My mission has been
to see every way fire can be conjured in its native setting and find out
more about its significance in other cultures. For as long as I can remember, I had wanted to see the fire saw in use by its true originators. With a
couple of weeks' leave from work, I headed to the islands of Luzon and
Mindoro in the Philippines, with my girlfriend Angelica, who is Filipina.
Here, the most common fire saw is made from bamboo.

Mindoro is the seventh-largest island in the Philippines, and
Mangyan is the collective term for the eight indigenous sub-groups that live
there. This time we did not need to trek for miles into very remote territory,
but found tribal groups living relatively close to civilisation. (Some tribes
on this island do still live more remotely and tend to run away at the sight
of somebody wearing Western clothes.) The majority of Mangyans live in
the lowland areas, and earn money from grass cutting, fruit picking, palay
harvesting, weaving baskets and other tasks. They enjoy simple village life
and are both hospitable and welcoming.

The village we visited had a community hut at the entrance, made
traditionally from bamboo and palm leaves – as was all the village. In
this hut, many men and women were making the most beautiful baskets
from bush materials – vines and other natural fibres from the forest – to
sell. I was astonished when I saw a little girl, no more than four years old,
making one of these baskets next to her father, who was doing the same.

She deftly and meticulously wove an intricate pattern, a practice passed down the generations. My enthusiasm for their basketry must have been apparent: when I left, they gave me a basketry awl made from an old hammered and sharpened nail fitted into a simple wooden handle.

I was there to find out about the fire saw, so I asked if anyone could show me how to use it. Most of them couldn't speak Tagalog – a common Filipino language – so my girlfriend could not communicate clearly with them. Somehow the message got across, and while the younger members of the tribe clearly did not make fire by friction (being so close to a town, where lighters and matches were available), they asked an older member of the village.

When we arrived at his hut, the older man appeared to be sleeping on a woven mat on the floor. There were a few baskets on shelves and hanging from the walls, and a couple of machetes suspended from the door in their wooden scabbards. Tribespeople live in a different way to westerners, but their houses are still very clean and you must always remove your shoes before you enter. Women are always sweeping the ground outside with long brushes made from the mid-ribs of coconut palm leaves. The sweeping sounds from the stiff bristles are a familiar background noise in these parts.

The young man escorting us called out a name and the older man stood up. A loud discussion in their native tongue ensued. I guessed the older man was about seventy or eighty. His age showed in his greying hair and beard and dark, leathery face, pitted with deep wrinkles and age spots. He broke into a gap-toothed grin, grabbed his machete and bounded out of the hut in the direction of the nearby forest. We heard the distinct noise of chopping. Within two minutes he reappeared, carrying a stem of bamboo. He started making a fire set outside his hut, using a bamboo fire saw. He used the machete with great expertise. The way he worked showed he had really lived these skills; it was clear that being able to use a machete to make fire had been the difference between life and death when he was younger.

Very soon, the fire set was complete, and the man quickly made a fire. What amazed me most of all was his stamina. This technique requires endurance and strength to achieve the vigorous action required, but he had no problem at all with it; his muscles flexed like those of a much younger man as he worked. He made it look effortless. There was a glint of satisfaction in his eye as the flames licked upwards and the fire started to give off heat. I asked him when he learnt the technique. He said that when he was a small boy his father taught him, because they lived a nomadic lifestyle deep in the forest, before they settled near the coast. In the old days, they always had to make fire like this, because matches and lighters were not readily available and expensive.

I could tell that the younger man had never seen this before; his eyes widened as the man sawed and smoke appeared. I looked around me; several more people, including some women and smaller children, had gathered to see what all the fuss was about, chatting animatedly to one another. Some started to pick up spare pieces of bamboo from the ground and experiment for themselves, trying to imitate the older man. The children laughed as they sawed. I hoped that my request would mean that this skill would be reignited in the village, if only for a short time.

## USE OF THE FIRE SAW

Several techniques employ a sawing motion to make fire. Each differs slightly from the other according to the region in which they are used, and ultimately by the materials available. Essentially, they all share the same principle: one piece of wood is rubbed across another at 90 degrees to the grain, in order to produce an ember. This is the method of making fire by friction that a novice is most likely to succeed with the fastest.

There are many ways the sawing principle has been applied, but they can be roughly divided into three main types. One relies on two pieces of stiff bamboo (one being the hearth and the other the saw). Another commonly employs the use of a solid piece of wood as the hearth and a separate piece of rigid hardwood for the saw. The third variation similarly employs a solid piece of wood as the hearth, but the saw is a flexible one, usually either made from a thin length of rattan or split out from a section of bamboo, and is used in a different manner to the former two variations. Because of this, I describe the flexible saw, also known as the fire thong, in its own chapter, Chapter 6.

OPPOSITE Mangyan tribesman coaxes an ember into flame, Mindoro island, Philippines.

The fire saw's traditional range is primarily restricted to tropical rainforest, but it is found outside those latitudes as well. Its western extremity is amongst the Chittagong Hill Tribes of Bangladesh and further north into the state of Assam in India, as well as on the isolated Nicobar Islands. Some accounts even record it being encountered in what is now Gabon. Moving east, it is used amongst the Karen people of south-east Myanmar, some Orang Asli (a collective term for eighteen ethnic groups who are generally considered to be peninsular Malaysia's original inhabitants) groups on the Malay peninsula, most of the tribes in the Philippines, and in a scattering of places across the Indonesian archipelago, all the way to the deserts of Australia.

## THE BAMBOO FIRE SAW

Sir Alfred Russell Wallace records the technique of the bamboo fire saw in his remarkable journal *The Malay Archipelago*, widely considered to be one of the most extraordinary travelogues of the Victorian age. This method of fire-making is still habitually employed across Indonesia and the Philippines today.

Bamboo occurs in tropical forests and warm, temperate regions around the world, and although it can sometimes be difficult to find, it usually grows in abundance. In these areas, it can be used in a quick and easy way to produce fire, providing the means to cut it is to hand. Most often this will be a large jungle knife of some sort, but all that is really

Bamboo fire saw set, Mindoro island, Philippines.

necessary is a tool that can split and scrape. A knife is not essential, and in an emergency a tool can be improvised. In addition, all the parts that are needed can be split out from a single, short section of dead, dry bamboo, meaning you only need to search for one stem. Bamboo has a coating of waterproof lacquer on the outside, which ensures the wood fibres underneath it are kept dry, even during the heaviest of jungle downpours.

### HOW TO MAKE A BAMBOO FIRE SAW SET

Any species of bamboo, providing it is in the correct condition and of the correct dimensions, is suitable for this technique. If you want to try this technique at home and bamboo does not grow nearby, you might be able to purchase some locally.

I    Collect a metre-long stem of dead, dry bamboo that is at least 5cm thick and has a wall no thinner than 4mm. When you look for bamboo, you will find a mix of living and dead wood. Dead wood rots quickly in the dampness of the jungle, so be sure the piece you select has retained its integrity.

II    Take this piece of bamboo to the place you want to start a fire; if it is raining, you should complete the process under a tarpaulin or improvised shelter.

III    From the stem you have collected, cut out a section that is at least 50cm long and free of nodes for at least 30cm, saving the other half for making tinder and kindling later.

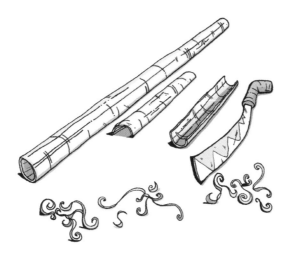

||||    There will probably be a node or two in the section you have removed, and you should ensure they are at the ends rather than in the space between. Split this section exactly in half by standing it upright, placing your jungle knife on top and tapping the back of your knife with another stick.

||||     Put one of these halves safely aside – this will become the hearth.

|||| |    Take the other half and split it in half again – one of these will become the saw.

### THE HEARTH

|    Take the piece you put aside and lay it down crossways in front of you with the inside of the stem facing the ground.

||     On the curved outer side, slice a narrow groove across the bamboo at 90 degrees to the grain. Take your time, as it is very easy to slip on the shiny exterior of the bamboo and cut your other hand with the knife. This would have grave consequences in the jungle.

Notching
the hearth

|||     Keep slicing until the knife is a hair's breadth from breaking through. At this point, use the tip of the knife to lance a small hole through the remaining wood. It is important that this hole is not made too big – make it 3mm long and 1mm wide.

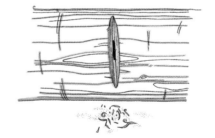

||||     Now, find the section you put aside for tinder and kindling right at the beginning, and scrape the outside, holding your knife with both hands. You should quickly produce a mass of thin shavings. Try to scrape back and forth in a way that leaves the shavings attached to the stem in two heaps of tinder, one at each end of the scraping stroke.

OPPOSITE Aeta tribesman scrapes tinder from bamboo, for use with his fire saw,
Zambales Province, Luzon, Philippines.

Adding retaining
splints to hold
the tinder

‖‖    When these are both the size of a chicken's egg, stop scraping and place them snugly beside each other on the inside of the hearth, one either side of the tiny hole.

‖‖ |    The final job is to take a pencil-sized piece of split bamboo or any stick, bend it in half gently so it only partially breaks, and place it inside the hearth, on top of the bundles of shavings so as to hold them in place and prevent them falling out. This is the hearth complete.

I have also seen some indigenous people carve an extra, very shallow groove on the inside of the hearth if the bamboo is particularly thick walled, but this is not usually necessary. When I have seen this extra groove added, the resulting splinters of wood were left attached at one end and the tinder was tucked conveniently under them.

### THE SAW

The sawing component is much simpler to prepare. Simply take the piece you put aside earlier and sharpen one of the edges by carefully shaving it down with your knife. There is no need to make it really thin and sharp; reduce the sides just enough to ensure the edge will fit into the groove in the hearth and make contact with the tiny hole.

### KINDLING

Before you attempt to make fire, split a handful of very thin lengths of bamboo from a spare section and put them aside somewhere dry. These prepared splints can be put straight on top of the tinder when it bursts into flame later.

The sawing edge must sit in the notch in the hearth while the two components are rubbed vigorously back and forth. The aim is that the friction dust will be pushed through the hole by the edge of the saw and collect on the inside of the hearth against the tinder. There are two ways to achieve this depending on whether you are on your own or working with someone else.

## SOLO

When making fire solo, firmly secure the saw edge-up at a convenient angle and take the hearth to it, holding it with both hands. The best way I have seen to accomplish this is as the Mangyan showed me in Mindoro (see OVERLEAF) which I will share here.

|     Cut a 5cm-thick sapling down with a machete, leaving about 30cm sticking out of the ground.

||     Split the top of this stem just enough to allow one end of the saw component to be tapped into it and held tightly. The other end is allowed to rest on the ground. By arranging it in this manner, the saw is held securely and at a suitable angle of about 45 degrees. When learning this skill at home, to save cutting anything you can hammer a dead stick firmly into the ground to act as the sapling.

|||     Next, pick up the hearth, checking the two tinder bundles are still positioned over the hole and are snugly packed in. Place the hearth onto the saw so the edge sits in the notch and begin to move it backwards and forwards.

||||     Start slowly, with a stroke that is about 20cm long. Once the set has warmed up and is producing smoke, gradually increase your speed and pressure.

|||||     After about twenty seconds, you should shorten the stroke to roughly 15cm. Intensify your efforts and give it your all for ten more seconds. The process should crescendo and cease on a peak.

卌 ||    When you have stopped, turn the hearth around and immediately blow through the hole. Don't be disheartened if you can't see anything glowing inside; the tell-tale sign you have been successful is a stream of smoke exiting the back of the hearth when you blow through the hole. Sometimes the hole will be clogged with friction dust. If this is the case, swiftly clear it with a thin splinter of bamboo. A convenient place to store this pre-prepared splinter is skewered amongst your hair in the same way the Orang Asli store a spare dart for their blowpipes.

卌 |||    Gently tilt the hearth over and allow the tinder bundles to rest on your hand. Remove the hearth slowly and look for the ember. It should be resting on top of the tinder. Put the hearth down carefully, wrap the ember up in the shavings and continue to blow until you have flames. Once this has been achieved, place the prepared kindling on top. When that has caught, add any other spare pieces of bamboo, and even the fire set itself if you wish.

### TANDEM

If there are two of you, you can work together (see LEFT). This time, the hearth is placed on the ground and held steady with the notch facing up, and the saw is held and moved. One person can hold the hearth steady while the other uses both hands to work the saw back and forth, or both people can have one hand on the saw and the other on the hearth in order to share the workload of the sawing movement. The rest of the process is the same as before.

### SOLID WOOD FIRE SAWS

The fire saw was made differently if people did not have access to bamboo, with the characteristic hollow stem that lends itself to the above technique. There were several variations. The sawing component, on the other hand, remained essentially the same: a sharp-edged length of wood, either the same hardness as the wood used for the hearth or slightly harder.

In Australia, while the hand drill was widely used, in many areas Aborigines relied on the fire saw. There were three main types: spear thrower and shield, natural split, and cleft stick.

OPPOSITE TOP Mangyan tribesman using the fire saw solo,
supported in the split stump of a sapling, Mindoro island, Philippines.
BOTTOM Aeta tribesmen use the saw in tandem, with the hearth held
against the ground, Zambales Province, Luzon, Philippines.

## SPEAR THROWER & SHIELD

In the central Australian desert, a 'miru' spear-thrower and an 'alkuta' shield, both usually made from mulga wood, were two items that warriors would carry with them into the bush. They had multiple uses. To make fire, a groove 5mm deep, 5mm wide and 10cm long was first carved into the surface of the shield. The edge of a spear-thrower was then sawn to and fro over the groove at 90 degrees to it, either as a pair or solo. The resulting friction dust would collect neatly in the groove and form into an ember before being transferred to a bundle of spinifex grass and blown to life. The next time the warriors needed to make fire they would saw across the same groove, alongside the previous scorch mark.

## NATURAL SPLIT

Cracks and splits occur naturally in old, dry, fallen wood, especially in arid climes such as the Australian desert, where the sun bakes everything. Aborigines would sometimes saw across a 5mm-wide split with the edge of a spear-thrower to make an ember. Just before the sawing commenced, dry, powdery kangaroo dung was dropped into the split to help with the production of the ember (the saw being worked over the top of it).

## CLEFT STICK

For the cleft-stick fire saw, the hearth is formed from a short, softwood branch that is split in half at one end and held slightly open with a small wooden wedge or a stone. Branches of varying thicknesses were employed, but the best is one about as thick as a man's wrist. A tinder bundle was placed on the ground with the split stick on top and the hardwood saw worked across the split where it was 5mm wide.

## SUITABLE WOODS

As for the bow drill (see page 57), as well as ironwood and mulga.

## FIRE CREATION STORIES & LEGENDS

Legends and myths about the creation of fire abound in almost every race and culture. Stories trying to make sense of the immensity and monumental scale of the universe, nature, seasons and weather have passed down through hundreds of thousands of years of storytelling from one generation to the next. In both the ancient and modern worlds, fire festivals and rituals reflect man's recognition of the natural world.

Fire has been important to mankind since the beginning of time, but how fire was created is shrouded in mystery and forms the centre of many stories. Studies show that the control of fire dates back 400,000 years. The general consensus is that at this point, the evolution of man begins, when humans became dominant above all others, as creatures capable of profound intellect and abstract thought. Ignoring other simpler scientific explanations, some stories hold that fire was gifted to man from the gods, whilst in others fire was stolen by gods, man or animals out of pity. In still other cultures, the creation of fire by friction led to its connection with sexuality; in some legends, fire's origin is traced back to the sexual act of animals or mythical creatures.

## PROMETHEUS STEALS FIRE

According to Greek mythology, the god Prometheus – who was known for his intelligence and wit – was tasked with forming man from water and earth. He did this, but became fond of his human friends and felt sorry for them. He stole a sacred lightning bolt from Zeus, concealed in a hollow stalk of fennel, brought it to earth and gave it to mankind. This was a precious gift, and he knew he would be punished. Zeus was furious and Prometheus was chained on Mount Caucasus, to have an eagle eat his liver forever.

## THE GRANDMOTHER SPIDER BRINGS LIGHT

A Cherokee legend tells of how Grandmother Spider brought light to the world. In ancient times, the sun was on the other side of the world, so everything was dark. The animals, who kept bumping into each other, decided that someone must steal the light. Both buzzard and possum tried, but ended up burning their feathers and tail. Using her eight legs, Grandmother Spider made a bowl of clay and wove a web towards the sun. There she put the sun in her bowl and rolled it home, bringing light and sunshine to the world.

## THE FALLEN ANGELS & AZAZEL INSTRUCT HUMANITY

In the ancient Jewish religious work the *Book of Enoch*, two hundred fallen angels and their leader Azazel descended from a heavenly realm on the summit of Mount Hermon. They were smitten with humanity, and using their material bodies, had sex with the humans. The fallen angels taught their new wives and children skills, magical knowledge and occult wisdom, and how to use tools and fire.

## RABBIT TAKES FIRE FROM THE WEASELS

A Native American tale tells of how the weasels were the only creatures with fire, after thunderbirds sent their lightning to the sycamore tree. They would not give any of it away, but the other animals could see the tree smoking. Every night the weasels would make a big fire and dance around it. The rabbit was the one animal brave enough to try to steal the fire, and the weasels welcomed him because they had heard he was a good dancer. He danced so close to the flames that the pine tar in his hair caught fire, and he ran off with it. The weasels called on the thunder-birds to make it rain to extinguish the fire, and it rained for three days. They were sure the rabbit could have no fire left – but he had built his fire in a hollow tree, and when the rain had stopped, he brought it out and gave it to the people.

## MAUI & THE MUD HENS

In Polynesian myth, the hero Maui was fishing in a lagoon with his brothers, when he looked back at the mountain and saw a fire burning. Man had not had a fire for many generations. He raced back to see a family of mud hens stamping out the embers. He decided to wait on the shore, and sent his brothers out fishing with a life-sized doll of himself. As the birds began to make their fire, he grabbed a hen – but realised he couldn't kill her, or the secret would be lost. The hen taught him to start a fire by rubbing sticks together, and Maui taught his fellow humans the secret of fire-making.

# CHAPTER 6

# FIRE THONG

**NEW GUINEA WAS ALWAYS A DREAM DESTINATION OF MINE. I LONGED**
to discover more about how some of the people there conjure fire, using
a fire thong or 'tekan'. I was fascinated by the country, its people and
environment. Most of all, I wanted to visit the mysterious jungle tribes
in south-eastern Papua. The Korowai tribe's first contact with the outside
world was in the 1970s when missionaries discovered a group of people
living in distinctive timber treehouses perched over 40m high above the
jungle canopy and still using stone tools.

Today, not much has changed; they still refer to Westerners
as 'ghost demons' because they are so rarely seen. The tribe are said to
frequently feud, and some are even believed to practise cannibalism. This
is one of the most remote places on earth, and I knew that this experience
to see another tribe – the Dani tribe – was my gateway to the indigenous
tribal culture. A trip further south to see the Korowai would take some
real knowledge of the people and the area.

I had read as much as I could about the area and the Dani tribe,
before finally setting off on my adventure, this time with my younger
brother Ben in tow. He had the same adventurous, indomitable spirit.
When I had told him that I was going, his face had lit up, and I suggested
he tag along for the ride. We flew from London to Jakarta, and from there
to Jayapura and onwards to Wamena. Situated at an altitude of 1,550m in
the heart of the Cyclops Mountains, Wamena is the only urban area in the
valley. The Dani sell their handicrafts here at the local markets, receiving
rupiah in exchange, but they remain poor.

The main area in which the Dani live – the Baliem Valley – was
thought to be uninhabited by people until as late as the 1930s, when in
the midst of a reconnaissance mission American philanthropist Richard
Archbold happened to notice a deep valley situated amongst tall mountains and clearly recognisable fields, an unmistakable sign of civilisation.
Despite some encroachment from the modern world since then, the Dani's
traditional way of life has proved remarkably resilient. The majority still

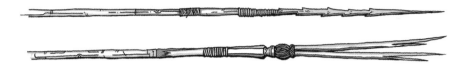

live the same way they always have, mainly due to the remoteness of their location. The bow and arrow is a common sight, with many still using them to hunt birds and small mammals or to dispatch the pigs they keep. The men still know how to make stone axes and adzes (a tool similar to an axe, which dates back to the Stone Age). Although tribal warfare has long been a part of their culture, the Dani are mainly a gentle, friendly, agricultural people who tend crops of sweet potato, coffee and tobacco between the peaks of New Guinea's staggering highlands, which rise to over 3,600m. Snow and ice can be found here in the form of glaciers all year round. The term 'equatorial ice' is a very fitting one.

The day after our arrival, we went to load up our bags with eggs, noodles, bananas and rice in Wamena's bustling market, so that we had food both for ourselves and to offer the village chiefs. The market was bustling with people going about their daily business – most having walked barefoot for many miles through the mountains, their beautiful string bags laden with all sorts of things and slung over their foreheads.

Soon Ben and I found ourselves clinging onto our bags and each other as we bumped along a dirt track up into the mountains beside our guide, Jonas. As soon as it became clear that the vehicle could go no further because the ground was impassable, Jonas hopped down and marched off at quite a pace, leaving Ben and me in his wake. The walk took us through a staggering landscape, with enormous peaks far off in the distance and rolling hillsides sectioned off into little terraces of crops, sweet potatoes, coffee and tobacco forming vast swathes of countryside.

We first passed a Dani village elder on one of the high trails up in the hills. He was completely naked apart from the traditional attire of a penis gourd and a headband of cassowary feathers. Despite having the face of someone much older, he boasted an impressive six-pack and lean leg muscles.

'Cigarette?' I took a box of clove cigarettes from my pocket and offered one to him.

He grinned, showing a mouth full of gaps and accepted the cigarette, before showing me a boulder to rest on. He muttered something to Jonas, who relayed the message: 'Would we like to spend the night in his village?'

'Yes, please. We would love to,' I relayed back.

We made the short walk to the village: a collection of several family huts, or to use the traditional name, 'honai', nestled in a valley. It was a great privilege to be welcomed in. Some honai are occupied by men and some only by women, and the men do not sleep in the same house as the women. Some honai are long and rectangular; these are usually used as the kitchen. This is where we were housed during the visit. Our honai had a simple but comfortable flooring of dry grass, and a central fire pit with two 2·5cm-thick metal bars spanning over the top. This formed a rack wide enough to suspend various cooking vessels over the flames. Several bundles of firewood were stacked neatly in the corner, each lashed with withies. The firewood the Dani used was very dry. It was remarkable in that a relatively thick splint could be lit easily with a single match, rather reminiscent of the resinous pine wood of the boreal forest – except it burned without a noticeable aroma. The honai were surprisingly clear of smoke; this tended to filter up through the tight thatch of the roof. This was always an amazing sight: looking down on a village nestled amongst the hills, and watching the smoke emanate from within, rising slowly up to form a blue haze, colouring the mountainside behind.

I spent a wonderful evening with my new hosts listening to conversations in Dani and Bahasa Indonesian. I slept that night on the grass floor of the hut. By 6.30 a.m., I had peeled back my blanket and started to warm my hands next to the cheery flames. We were sitting cross-legged on the grass floor listening to the sound of the kettle lid jumping and rattling. This is always a pleasant sound when staying out in the bush, especially on cool mornings. The smell of coffee soon filled the air as Jonas tended to some eggs which were dancing around in a pan of water.

That morning, I asked my host if he would show me his tribe's traditional way of making fire. I waited eagerly as Jonas translated, and I tried to interpret the thoughtful expression of my Dani host as the message was communicated. There was a bit of chatter between the two of them. Suddenly my host took a long final draw on his cigarette and exhaled a stream of smoke towards the roof before leaping up onto his feet. He dashed off, and it wasn't long before he returned holding a long, thin strip of rattan as well as a thicker piece of wood. I witnessed a scene that has remained unchanged for countless centuries. Using two pieces of dry rattan, he produced fire in less than a minute. Then he cut a new length of rattan and offered it to me. He obviously wanted to see me attempt it.

I tried to emulate his body position as closely as I could – I know from many years of experience that success with these things is in the finest detail, so I paid close attention. I began to saw the thin strip of rattan back and forth as he had shown me. I started quite steadily and smoke appeared as usual after only a few seconds of friction. At that point, I increased the speed of my hands and went as fast as I could. I kept going, knowing that with this technique, you should stop only when the rattan thong snaps. The whole thing took about fifteen seconds. Once it had snapped I immediately knelt down and inspected the split in the hearth

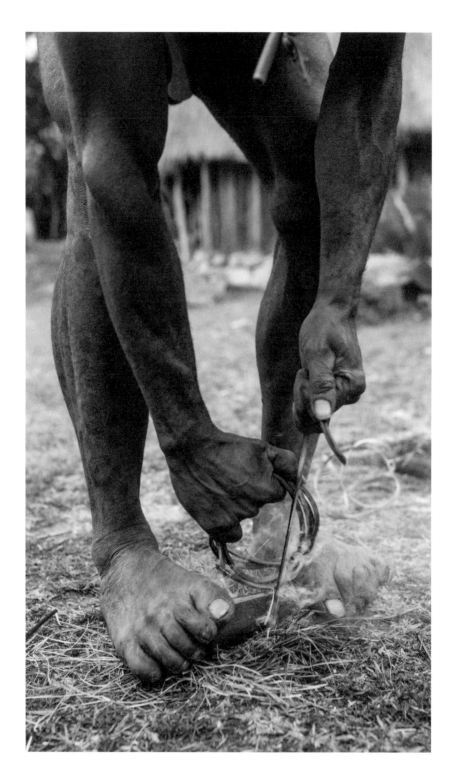

stick. Success! The split was full of charred dust, which continued to smoke on its own. As I gently blew some air onto it, I saw the orange glow of an ember. At this point, another man came over and sat down with us. The two of them swapped a few words in their language. They were both smiling, and one helped me coax the ember onto a bed of dead, dry grass, which I wrapped up and blew to life. Laughter erupted, and I felt a heavy pat on the back. We shook hands and watched as the flames from the grass died down and the leftover ash tumbled away in the wind.

It is no exaggeration to state that visiting this great island felt like journeying to another planet. It seemed so cut off from the rest of the world. Somehow, though, as I have experienced in many other remote places around the world, possessing a knowledge and appreciation of nature, and being able to look after oneself in the bush, breaks down language and cultural differences and brings you closer to the local people. You have something in common and something to share. You gain their respect and develop deeper relationships. No other element brings people closer together than fire. It is the magic ingredient that makes us human and one that allows us to connect with each other regardless of who we are.

## USE OF THE FIRE THONG

The fire thong is a variation of the fire saw, but the sawing component in this case is made from a flexible length of rattan or bamboo instead of a stiff, blade-like length of wood. Its traditional range is not greatly dissimilar to the other sawing methods, and it is sometimes found being used alongside them. It is used by the Naga people in the Naga Hills of north-eastern India and north-western Myanmar. On mainland New Guinea and some of the surrounding islands, this technique dominates and is relied upon by hundreds of tribes. Some groups of Orang Asli on the Malaysian peninsula use it too; I have encountered it amongst the Semai in particular.

In the Philippines it is used by the Bataks in central Palawan, some Mangyan groups in northern Mindoro and amongst the natives of Casiguran in north-eastern Luzon. Across this broad area, people often

**LEFT** Dani tribesman using the fire thong, Papua, Indonesia.

Fire kindled by Batak tribespeople with the fire thong, Palawan, Philippines.

carry a dry length of rattan with them as we might a box of matches. In many places, people wear armbands made of rattan. I have seen this in the highlands of New Guinea, as well as observing rattan waist-belts worn for the same purpose. The hearth, on the other hand, being easier to find, is quite often sourced from the bush whenever fire is needed, though some people in New Guinea carry the whole set with a long length of rattan curled up into a tight ring.

In Borneo, both the Kayan and Kenyah people used it ceremonially when naming a newborn child. A specially made hearth is created from a soft piece of wood which is beautifully carved into the form of 'Laki Pesong', the god of fire. The legs of the god form the hearth and a thong of dry but flexible bamboo is then sawn across them until it snaps – an ember being produced in the process. The two broken pieces are then inspected. If the cane has broken into two equal lengths, it is seen as bad luck and the name of the child is changed before being subject to the same test.

Quite unusually, there are accounts of this method in isolated areas far from this eastern region. South of the Ogooué river in what is today central Gabon, there is one account that the Bakalai people employed it, but the description is not quite thorough enough to differentiate it from the other sawing methods with absolute certainty.

Using the fire thong was not only limited to areas where rattan or bamboo occurred. All across Europe, the thong was sometimes employed for ceremonial purposes in the production of 'need-fire', a superstitious practice called upon for many different reasons. Up until the middle of the nineteenth century, it was still widely believed that if the same fire was used for an extended period, it would become stale and lose its mystical powers. Whole communities would extinguish the fires in their homes, and relight them from the embers shared out from one communal need-fire made by friction.

Need-fire was also produced at other times as a remedy for plague or disease in cattle or swine; cattle were sometimes driven between two large fires conjured in this manner in order to be fumigated by the sacred smoke. Even bewitched fishing nets and tackle, and orchards of fruit trees, were bathed in this smoke to cure and protect them. The resulting ashes from need-fires were sometimes spread on crops to protect them from vermin or used as a remedy for bodily complaints, the ashes being sprinkled onto the affected part or mixed with water and consumed.

From region to region, there were many different ways need-fire was produced. Strict rules and traditions dictated who was involved, and how and where the fire was made. In Scotland, the fire was sometimes made by eighty-one married men working on a small island in a river or on a knoll; in Serbia, it was a young boy and girl working in a dark room. Sometimes a crossroads in the highway was preferred.

In Sweden, there were several methods to producing a need-fire. In northern Sweden, people would sometimes form a tough ring from a withy of birch, and place this around a dry tree stump; a stout pole was threaded between the withy and the stump, and twisted once until it was taut. The operator would then proceed to move briskly around the tree while holding onto the pole to achieve friction, and ultimately combustion. Another method was to tie one end of a rope to an anchor point, the other end being pulled taut, with another person rubbing a hearth stick along this line. There is very little recorded evidence regarding how the friction dust collected in these cases, or whether it was even needed. One account suggests that linen, sometimes impregnated with tar, was placed between the moving parts used in some methods in order to capture the first ignition.

**STONE ADZE**
Dani tribe, Papua Province, Indonesia.

**FIRE THONG SET**

bound for travel with a strip of bark, Dani tribe, Papua Province, Indonesia.

## HOW TO MAKE A FIRE-THONG SET

Although there is some variation in the types of thong used, the greatest difference lies in the construction of the hearth piece. Rattan or bamboo are suitable woods for the thong; the hearth can also be made from these woods, as well as from most of the softer woods found in the tropics.

When practising outside the tropics, you can try either the rattan or bamboo you are using for the thong or any of the woods suitable for the drilling methods.

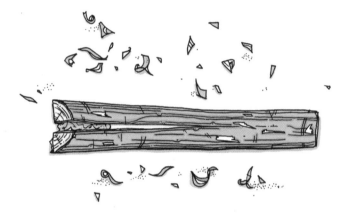

THE HEARTH

There are two main ways to make the hearth. Experiment with them: one is sometimes more effective than the others, depending on the materials you have to hand.

I encountered the most common style, the split-and-wedge hearth (ABOVE), being used by the Dani in the highlands of New Guinea and in the jungle of Palawan. A smooth, debarked, round branch about 55mm in thickness and 350mm long is split open down the middle from one end. The resulting opening is wedged at the very mouth of the split with a small, flat pebble. The split is allowed to run down the stick as far as is needed to allow the end, where the pebble sits, to be held open by 12mm.

The Semai people on the Malaysian peninsula use a funnel hearth (OPPOSITE) very effectively. It is easier for a beginner to succeed with because the ember forms in a more visible manner than with the split hearth arrangement.

|  Select a branch of the same dimensions as for the split hearth, and split it completely in two along its length. Only one half is needed to make fire – set the other aside.

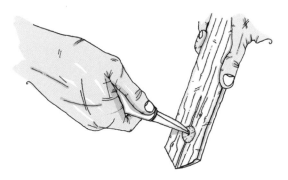

||  Using the tip of a knife, carve a funnel-shaped pit near one end of the board, starting from the flat side and working through to the round side until you are all the way through. The entrance and exit of this hole should be round, starting with a hole about 25mm wide, which tapers and gets narrower the deeper into the wood it gets. The exit hole should be about 5mm across.

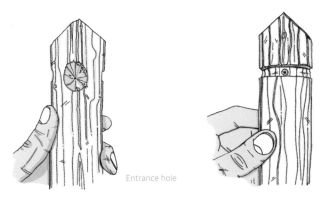

Entrance hole

Exit hole, with guide channel

|||  Carve a shallow channel that crosses the grain of the wood at 90 degrees. Ensure it crosses the exit hole exactly. This channel will act as a guide to keep the thong in the correct place, so it doesn't need to be anything special – just enough to guide the sawing action later.

Hearths are sometimes made from bamboo and are set up in the same way as those designed for use with the rigid bamboo saw (see Chapter 5).

**THE THONG**

There are two ways of making the thong. A whole piece of dry rattan 6mm in diameter, left in the round, can be collected and used immediately. If this cannot be found, a length of either rattan or bamboo can be split off from a large piece, and whittled down until it is 6mm wide and 4mm thick. Ensure that the polished outer layer remains on one of the 6mm wide sides.

## HOW TO MAKE FIRE USING A FIRE THONG

I     Place a bed of tinder on the ground and put the hearth stick on top of it.

II     Take the thong and thread it under the whole of the hearth board so it is sandwiched between it and the tinder, and the two ends stick up towards you. If you are using a thong that has been whittled from a large piece, ensure the polished outer side is closest to the ground.

III     Pin the hearth to the ground with both feet in a way that prevents it from wobbling. This will take a little getting used to as you try to find the most stable position. Try to emulate what you see in the photo (OPPOSITE).

IIII     If using a split-and-wedge hearth, ensure the thong crosses the split where it is 8mm wide. If using a funnel or bamboo hearth, ensure the thong is aligned exactly with the notch you cut earlier, and runs directly over it.

IIIII     Slowly begin drawing the thong back and forth around the hearth ensuring it continues to remain in the correct place. Start with your arms spread wide. As the thong heats up and becomes more flexible, you will be able to bring your hands closer together. If you start with your hands close together, you risk snapping the thong. Once you see smoke, increase your speed, and keep going until the thong snaps.

卌 |　Inspect the split in the hearth stick, while being very careful not to knock it or move it abruptly. The split should be clogged with dark friction dust which continues to smoke on its own. This is the precious ember. If you are using the funnel hearth instead, you will clearly see the friction dust as it gathers in the hole.

卌 ||　Gently blow onto this wood dust to encourage the tiny ember to grow. When you see it glowing orange, use a thin twig to push it out of the split or the hole and onto the tinder underneath as you simultaneously lift the hearth stick up. Place the hearth stick aside.

卌 |||　The process from now on is the same as the other ember-producing methods. Carefully wrap the tinder around the ember and blow it into flames. If you are using a bamboo hearth, follow the fire-making guidelines for the bamboo fire saw in Chapter 5. Apart from the use of a flexible sawing thong, the process is identical.

Fire thong in use, with tinder bundle ready in the foreground, Batek people, state of Pahang, Malaysia.

## A WIRE THONG & AMMUNITION PROPELLANT

An unusual modern technique which makes use of this flexible thong action can be employed in an emergency if you have access to ammunition and a length of fence wire. Cut open a shotgun cartridge, or carefully prise out a bullet head from its case and tip the propellant into one pile on some tinder.

|     Take an 80cm length of wire and fit each end with a toggle handle.

||     Vigorously work the length of wire back and forth around a dry branch in a similar way to the previously mentioned thong techniques.

|||     When it is smoking profusely and you are at the climax of your effort, immediately touch the hot part of the wire onto the propellant. It will burst into flame, lighting your tinder in the process. Be aware that the propellant will flare up a little when it is lit.

OPPOSITE A Dani warrior nurtures an ember into flames, Papua Province, Indonesia.

# CHAPTER 7

# FIRE PISTON

**SIPPING AN EARLY MORNING HOT CHOCOLATE IN HEATHROW'S**
departure lounge, I knew I was embarking on a trip that was probably
going to be a very small needle in a very large haystack, but it was a needle
I desperately wanted to find. My task: to find one of the last – if not the
last – makers of perhaps the most intriguing of all the fire-making techniques: the fire piston, or as it has also been described, the fire syringe.
It is a peculiar method, with roots in Europe, but one that has long been
depended on across south-east Asia, most notably in the Philippines and
peninsular Malaysia. Today it has all but died out.

It works through the same formula as the diesel engine, using the
principle of heating gas by rapid and adiabatic compression, and mixing
it with fuel. In a fire piston, if it is done quickly and efficiently it can
reach temperatures in excess of 425°C. In fact, Rudolph Diesel – the engineer and father of the engine named after him – was inspired when he
attended a presentation given by German scientist and engineer Carl Von
Linde at the Munich Technical University in the 1870s.

Von Linde had just returned home after spending many months
travelling on a lecture tour. He paused during the presentation, put a
cigarette to his lips and removed a device that had been gifted to him by
the people of Penang Island off the west coast of peninsular Malaysia – a
fire piston. With a sudden strike of his hand, he plunged and removed
the piston, and using a small pick to prise out the glowing tinder, used it
to light his cigarette. Diesel, one of Von Linde's most promising students,
had become frustrated with the low efficiency of the internal combustion
engine during his recent experiments. What he saw at this presentation got
him thinking. Instead of using a spark to ignite fuel, perhaps an engine
reliant on this compression principle could be made.

Although there have been many designs of fire piston throughout
history, commonly they consist of a short, narrow bore that is drilled into
a block of hard material into which a well-fitting piston, usually made
from the same material, is rapidly forced. The piston itself has a shallow

OPPOSITE The beautiful primary rainforest of Pahang state, Malaysia.

depression in the tip, into which a small quantity of tinder is packed. A gasket forming an airtight seal sits close behind this. The tinder is set smouldering when the piston is rammed smartly into the bore. As soon as this has taken place, the piston is swiftly removed to expose it to air, and the tinder, glowing like the tip of a cigarette, is carefully removed with a thin pick of wood and used for small jobs or, if a fire is required, transferred to a larger bundle of tinder before being blown into flames.

Researching the origins of the fire piston, I discovered there was an Orang Asli man who was one of the last indigenous individuals to possess the skill and knowledge of how to make the clever device by hand. Although I had scant information as to where he lived, I knew he was from the Semelai group and lived in peninsular Malaysia, somewhere near Tasik Bera, a lake in southern Pahang, the third-largest state in Malaysia. I set off optimistically...

Inspecting the map on the floor of my hotel room in Kuala Lumpur after my long flight, the nearest town of any decent size near to Tasik Bera was Temerloh. The following morning, my girlfriend and I jumped on a bus and travelled the two hours north-east. Our last hotel had stunk terribly of durian, a pungent, custardy fruit, best known for its fragrance which has variously been compared to rotting flesh, sewage or, at best, ripe cheese. To my relief, I noticed a 'no durian' sign hanging in the lobby. We booked in for the night, and using the few words I knew in Malay combined with a phrasebook, I asked the lady at reception if she had heard of Mr Jamri, the fire-piston maker. She looked completely blank. Was it my pitiful attempt at Malay or that she really didn't know who he was? Considering we were still 60km or so away from the lake where Mr Jamri supposedly lived, it was a bit of a long shot. I asked her if she would organise a taxi for us the

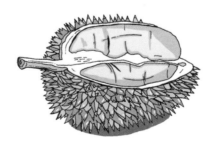

following morning, stressing that we would prefer an English-speaking driver if at all possible; this request seemed to be understood.

In the morning, the taxi driver turned up as planned. We had packed a few essentials – a machete, camera, water bottles, rice, cooking pan, sleeping mat and an InReach communication device, which allows for two-way communication and provides remote tracking. We weren't planning on going into the bush on foot, and had arranged to be back at the hotel that night, but I learnt very early in my career while travelling in remote places that it is always good to go prepared to spend a night or two away. Transport is never available when you need it, it gets dark quickly and you are usually in an unfamiliar environment. When you are prepared, however, this can be fun instead of an inconvenience or a danger.

'Selamat pagi! Apa khabar?' I said, a Malay greeting meaning, 'Good morning. How are you?'

'Baik!' came his reply: good.

'Do you speak English?' I enquired. He shook his head. Wondering what to do, I decided the best thing would be to show him some photos of fire pistons I had with me, mention Mr Jamri and see what his response was. He appeared interested when he looked at them and nodded as though he knew what they were. Great, I thought. He was a gentle and pleasant chap, and he seemed to be confident of where to go, so we jumped into his car. We headed south, initially following the Pahang river, before driving through endless oil-palm plantations. After a couple of hours, the driver pointed out a signpost for the Lake Bera resort. I realised that he didn't know the whereabouts of Mr Jamri – we had spent the past twenty minutes going round in a big circle.

I thought the resort might be a good place to have some iced tea and ask around for Mr Jamri, so we turned in and followed an unsealed track before coming to the 'resort'. A bulldozer was ripping the main building down, and there was no one around apart from the bare-chested driver, his head draped in an old T-shirt – the only thing providing him with shade. He turned the chugging engine off when he saw our car doors open and stepped down from the cab.

Our taxi driver quizzed him as to the whereabouts of Mr Jamri. When I showed him photos of a fire piston, he was none the wiser – and who can blame him? It was a bizarre request from a random, mad-sounding

Englishman – probably the last thing he expected. Suffice to say there was no iced tea either, but he did suggest we ask at an administrative centre just down the road.

Again our driver asked the guard at the entrance; again the man looked blank. Feeling increasingly deflated, we were directed towards another building but were met with further blank faces. After passing the photograph around all their colleagues, finally one person seemed to recognise the fire piston and we had our lead at last! I still don't know what they were actually doing in that office, but I was glad that we had met them. Our driver told us that the area Mr Jamri lived in was still an hour's ride away but we only had two or three hours of daylight left. Armed with this new information, we decided to head back to the hotel in Temerloh to get a good night's sleep and to give it another shot in the morning.

The taxi driver picked us up again early, and we spent a couple of hours on the road, until it became progressively smaller and rougher, and the oil-palm plantations turned into secondary forest. From there our driver had to wave down at least three or four locals whizzing past on mopeds to ask for directions. After making several twists and turns along bumpy, muddy tracks through beautiful forest, we finally arrived at a remote Orang Asli village.

As we entered the village, we found a tiny store selling tinned fish, cigarettes and cold(ish) drinks. Finally, some iced tea! Coincidentally, we also found the chief here, relaxing under an umbrella. He spoke a little English, and knew Mr Jamri (OPPOSITE). He understood the purpose of my visit; once we had finished our drinks, he accompanied us to Mr Jamri's house, which was only five minutes away. The chief opened the front door, called out, and a voice came from inside.

We were welcomed in, having removed our shoes. Mr Jamri was not an old man – perhaps fifty years of age – and he didn't seem at all disconcerted by two random people turning up on his doorstep. Despite its remote location, his house was not a traditional tribal design and was relatively modern, with a crinkly tin roof. I greeted him with a handshake, delighted to finally meet him.

His smiling wife brought us some hot tea as we sat cross-legged on the floor. I explained more about who we were and where we were from, and the chief explained to him why we had come. Mr Jamri looked

surprised, but then beamed broadly and disappeared into his back room, returning with a large selection of fire pistons. They were a variety of shapes and colours: some plain wood with the natural grain showing, others painted or varnished. Even though the chief knew a few English words, I couldn't ask him everything I wanted to. Mr Jamri demonstrated how the fire pistons were used, chatting animatedly – much to my frustration, as the taxi driver looked completely absorbed by his speech. He showed me the special tinder that is traditionally used with the fire pistons. He peeled back a leaf from the trunk of a caryota (or fishtail) palm and collected the fluffy 'scurf' where the base of the leaf met the trunk. I bought a piston from him, and he gave me a long length of tree-bark fibre in case I needed to replace the gasket when it was worn out. He also put a large handful of the caryota palm scurf into a carrier bag for me.

Mr Jamri told me, via the chief, that he had learned the art of making fire pistons from his father, who in turn had learned from his father when he was a child. It was obviously a long tradition in his family. Using a few basic hand tools – a machete, a small craft knife with a long handle known as a 'penat', and a hand-drill – Mr Jamri had produced the most perfectly functioning piston I have ever seen.

# HOW TO MAKE A TRADITIONAL FIRE PISTON

Making a fire piston from wood in the traditional manner is a fun but challenging task and suitable only for those with advanced carving skills. While it is a reliable device, it is much more difficult to construct in a survival situation, so would never be the first choice in these circumstances. The cylinder is simple enough to make but the piston requires neat and highly accurate work that is difficult to achieve if you are inexperienced.

It is best to choose a piece of dense, heavy wood with a tight, closed grain and ensure it is thoroughly dead and dry. Avoid wood that has split as it has dried. The technique relies upon the compression of air in the cylinder and so if the wood you choose has a grain that is too open, the air will be pushed out between the fibres. Bear in mind that it is even possible to blow through medium dense woods when they are dry – you can test this by smearing a little washing-up liquid at one end of a log and blowing hard into the grain at the other. Needless to say, avoid any wood that makes this possible. Collect a piece of wood that is 11cm long, no less than 8cm thick and still in the round. Saw this piece neatly so its ends are flat and 90 degrees to the sides. Both the cylinder and the piston will be made from this piece of wood.

## SUITABLE WOODS

Apple, beech, blackthorn, box, elm, hawthorn, holly, hornbeam, maple, oak, pear and yew. If none of these are available, select the hardest, dense wood you can find locally or visit a wood-turning suppliers.

## THE CYLINDER

You should work on the cylinder first because it is easier to adjust the piston diameter to fit the bore than to adjust the bore diameter to fit the piston.

I    Stand the piece of wood you have collected on a sturdy chopping block. Split it exactly in half, and drill a hole that is 8mm in diameter and 7·5cm deep into one of the halves.

II    Ensure the drill stays parallel to the sides of the wood and does not wobble as it works its way in. The hole should be drilled in a place that will ensure the core of the wood will not form any part of the finished cylinder. To do so is to risk the cylinder splitting when in use. Interestingly, in remote parts of south-east Asia, before metal drill bits were available, the bore was sometimes drilled over several days by hand using a wooden stick twirled between the palms. Some pistons were even made from particularly thick-walled species of bamboo which negated the need for any drilling at all.

|||     Having drilled the hole, make it smooth by fitting the end of a narrow strip of sandpaper into the end of a thin, split stick and wrapping the remainder of the strip several times around the stick – bringing the split closed in the process. This sanding tool can be pushed into the bore, and twisted and moved in and out to smooth the inside. Don't over-sand – the aim is to smooth the bore, not change the shape of it.

||||     Once complete, the sides of the block can be worked down to the finished dimensions. Start by splitting large chunks off. When you get to within 8mm or so of the edge of the hole, carefully smooth the whole thing with a knife. The finished diameter of the whole piece should be about 2·5–3cm. It can be decorated if desired.

### THE PISTON

|     Take the other half of the block you split earlier, place it on the chopping block, and split and carve it down until it is neatly rounded, and the same diameter as the cylinder.

||     Poke a thin stick into the bore in order to measure the depth of it as accurately as you can. Use that measurement to determine how long the stem of the piston needs to be; the piston needs to reach right to the bottom of the cylinder.

|||     Mark the measurement on the wood, then carefully saw all the way around the piece of wood in line with the mark, being careful to cut only a few millimetres deep.

||||    Take your time and carve the rest with a knife to ensure accuracy, and avoid going too deep. The finished diameter of the piston should be ever so slightly less than 8mm as it needs to slide easily in and out of the cylinder. Rest assured it will become airtight once a gasket is added to the stem later.

||||    When you get close to the finished diameter, use your knife in a scraping action to gradually smooth the surface. It is essential that the finished piston is absolutely straight, so keep an eye on this throughout the entire process.

|||| |    The piston is now almost complete, but there are two more little jobs needed to finish it off. Carve a narrow 1mm deep and 2mm wide trough all the way around the stem about 7mm from the tip. This recess will hold the gasket in place and prevent it from sliding up the stem when it is forced into the cylinder. Traditionally in the far east, the gasket is made by winding a thin fibre of bark from *Artocarpus odoratissimus* – a tropical tree commonly known as 'terap' in Malaysia or 'marang' in the Philippines – around the piston stem. Many other vegetal fibres also work. In Britain, I have had success with the fibres from lime, willow and stinging nettle, but as always, experiment with what is available to you. Even a thin piece of cotton from a rag will work. Modern fire pistons usually use a rubber ring to achieve the same job.

|||| ||    Whatever fibre you use, dunk the fibre in water until it is saturated, then wrap it around the stem in the recess a few times until you think it is bulky enough to fit snugly into the cylinder. Test it for a good seal. You need to strike a balance here; it should require a firm strike of the hand when plunged down, but not be too difficult or time-consuming to remove.

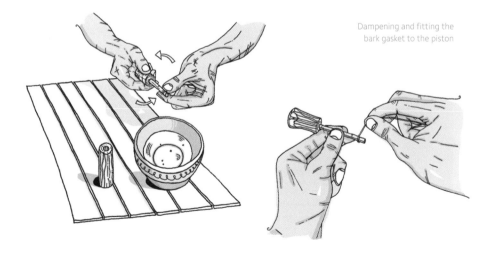

Dampening and fitting the bark gasket to the piston

129

||||| |||   The last job is to carve a cup-shaped depression about 3mm deep right on the very tip of the piston. Use a knife with a fine, sharp point to achieve this. This is where the tinder will be placed.

### TINDER

The finest tinder is required for this technique. The dry, fluffy scurf from a caryota palm (see BELOW), like that which Mr Jamri gave me, is traditionally used in the far east. Usually a leaf of the palm is peeled back away from the trunk so as to expose the area of trunk underneath the base of the leaf. Here is where the scurf can be found and gathered in abundance. It is usually dry, but if not, it must be made so before it can be used. The indigenous people collect this downy material and store it in a bamboo tinder box. Elsewhere, dry herbivore dung, cramp ball fungus, horse hoof fungus (and similar species), chaga and char cloth (a lightweight organic fabric that has been treated so it has a very low ignition temperature) can all be used.

# HOW TO MAKE FIRE
## WITH A TRADITIONAL FIRE PISTON

There is a knack to using a fire piston, and some can be tricky until you become accustomed to them. I have found the native Malaysian design described above both easier and more reliable to use than the highly engineered, machine-made versions available commercially.

I Place some tinder in the depression in the tip of the piston – this is most easily achieved with a toothpick. Pack plenty in and leave a small amount protruding very slightly.

II Dip the tips of your finger and thumb into water and apply some to the fibres forming the gasket. This will cause the fibres to swell which is crucial in order to achieve an airtight seal. Take your time and apply plenty of water, being careful not to make the tinder damp. Use your thumbnail to bunch the fibres up together a little if needed.

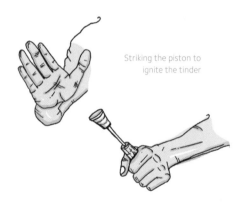

Striking the piston to
ignite the tinder

III Insert the piston into the cylinder by about 1cm or until the fibres are just inside. Holding the cylinder in one hand, strike the top of the piston with the other hand with a committed blow, forcing it right to the bottom.

IIII As soon as you feel it come to rest, immediately remove the piston as fast as you can. The whole thing should be as though it is one movement. If the removal is not fast enough, the tinder will use up all the oxygen inside and will go out.

IIIII Once the piston has been removed, immediately blow gently on the tinder to encourage it. You should see it glowing like the tip of a cigarette. Using the toothpick, swiftly but carefully prise out the smouldering tinder and place it in contact with more of the same tinder to extend the life of the ember.

IIIII I Now you can either use the ember for a quick job – such as lighting a well-deserved Cuban cigar to celebrate – or if a fire is required, transfer it to a larger tinder bundle and blow it to flames.

CHAPTER 8

# SPARKS

**DESCENDING DOWN A METAL LADDER, CAREFULLY PLACING ONE FOOT** behind the other, the temperature dropped noticeably and the jet-black flint around me shone with artificial light. Ben was ahead of me and my dad behind me. We were heading down the shaft at Grime's Graves, a fascinating Neolithic site, in open heath countryside near Thetford Forest in Norfolk. I was only about seven or eight and I was instantly captivated by the sights around me.

Despite its eerie feel and name, Grime's Graves is not a burial site, but the only Neolithic mine open to visitors in Britain. It is the most studied site of its type. It seemed incredible as we stepped down towards the floor of the mine that it was dug between 2200 and 2500 BC, using deer antlers for picks and animal shoulder-blades for shovels. These mines are considered the earliest sites created in Neolithic landscapes, with flint mining one of the first pursuits carried out by the growing Neolithic communities at that time. On one of the antler picks discovered by archaeologists on the site a miner's fingerprint was identified – still in place after 4,000 years.

From the air, the area has a strange, almost lunar appearance due to the series of sunken pits – remnants of more than 400 vertical mine shafts that were dug down through the chalk. We had walked across the dips and humps of earth to get to the mine and it seemed quite incredible that many were 14m deep. To think about the manpower involved is almost impossible. They were dug to reach what is known as

'floorstone' – a layer of very high-quality flint that was at first used for the production of stone tools, and later, with the arrival of metal, for gun flints and building materials. Once the miners struck the floorstone, horizontal galleries and tunnels were dug to follow the flint layer.

Once we emerged back at ground level, we caught sight of our guide, who was dressed crudely in old animal skins. Although this was more than likely historically inaccurate, it really conjured up a Stone Age atmosphere. I was buzzing with excitement to find out more. Around him the floor was strewn with flakes of flint on old blankets, and items such as antlers, flint daggers, spears, bows and arrows, and animal skins. On the wall in the mine there had been a flint axe that I couldn't keep my eyes off – it was so beautifully made. There has been speculation that some items such as flint axes had cultural significance – many that were found were buried in hoards and had never been used.

Our guide demonstrated how to make flint arrowheads, first by knocking a broad, thin flake off a 'core' with a hammer stone. He showed us how the finer shaping could be administered with the thin, pointed end of a deer's antler. The finished article was really intricate, and it was handed around the group before being given to someone to keep.

Our guide showed us how people in the Stone Age used to conjure fire. He had a lump of iron pyrites that had been split open to reveal a beautiful starburst pattern inside. Kneeling down with this stone in his hand, he proceeded to strike its surface with a small piece of dark flint.

As he did so, I could just about see some tiny, dull, red sparks floating down after each impact. He struck the two pieces together a few centimetres above some tinder placed on the floor. The tinder was a chamois-like slice of horse hoof fungus, a small section of which had been scraped up into a little fluffy pile – this is where he said he wanted the spark to land. Most of the sparks, though, landed around it and went out immediately. After twenty or so strikes, a tiny wisp of smoke emanated from the little pile; a spark had made a direct hit. Now there was an orange glow as the tinder smouldered and the ember grew. He put it into a bundle of dry honeysuckle bark and blew it into life.

Getting home later that day, I couldn't wait to smash open some flint. Of course, I also wanted one of those 'magic firestones' that I had seen at the demonstration. Where could I find a piece of iron pyrites? I asked my parents. My dad, who is a jeweller, told me that you could also make sparks with carbon steel if it was struck on flint. He said he would check in his workshop the next day for a spare piece of steel. The following evening, when I heard the jangling of his keys in the door I ran up to him. Pre-empting my eagerness, he opened his hand to reveal a 5 × 5cm square of thin carbon steel. Of course, I was straight out of the back door to find a random shard of razor-sharp flint that I had scattered on the patio a few hours before. Unaware of any proper flint-knapping techniques, but desperate to make arrowheads and knives, I used to find nodules of flint that had been turned over in the fields by the plough, bring them home and throw them at the ground until they smashed. I struck the steel against the flint, and was amazed at the resulting sparks. It wasn't exactly as I had seen at Grime's Graves but it was still astonishing to me back then. Even today, I see grown men on my courses absolutely over the moon when they first make fire with flint and steel. My search for a nugget of iron pyrites continued for quite some time afterwards.

All around the world, our ability to create sparks through the act of percussion has been utilised to make fire. For thousands of years, people have relied upon this ability. When we undertake expeditions in remote areas today, of all the fire-making gadgets we can carry with us it is a modern spark-making device that offers the greatest all-round reliability.

In every instance that I have observed percussion being used by tribespeople, it has always been just one in a repertoire of techniques known to them. This is not meant to imply that percussive – or, for that matter, friction – methods require backup. It is, however, an observation that serves well as a lesson for any wilderness traveller: always take the proper precautions, and have more than one method of making fire with you. If a Kalahari bushman of southern Africa was to empty out his hunting bag and quiver, you would probably find a hand-drill set, some matches or a lighter, and a piece of steel along with a couple of stones for making sparks. People living in the bush know only too well how important it is to be prepared.

It seems plausible to me that the knowledge of how to produce sparks originated as a chance observation during the manufacture of stone tools. Iron pyrites struck together with a glassy rock was a widespread combination used to produce the earliest sparks, but how might this coupling have arisen? Flint, quartz, chert, obsidian and other glassy rocks were an extremely useful resource, providing the material needed for making the everyday cutting tools of the past, but in order to be fashioned, these rocks had to be struck with other substances that could fracture them. Why iron pyrites? Relative to its size, it is heavier than most other rocks. That may have been an attractive quality to our early ancestors looking for hammerstones to use in the production of stone tools. Although iron pyrites is in fact brittle and completely unsuitable for this use, this is likely to have been discovered only while it was being tested in this manner – and by then, of course, sparks would have been flying.

Despite the advent of matches and lighters, the popularity of making fire with sparks has continued – mainly being used by those venturing out into wilderness areas. When we are close to civilisation, we can replace faulty lighters and wet matches very easily, but in the bush things are different. As always, we must take heed of the 'KISS' principle –

Keep It Simple, Stupid. Lighters have small moving parts and a limited fuel supply; matches are easily ruined by moisture, take time to dry out, and are limited in quantity. Although both are very convenient and get the most frequent use on my personal trips into the wilderness, I would think very hard before setting foot into such areas without making sure I had a modern ferrocerium rod in my kit.

There are several methods for producing sparks. Today some remain in use for serious bush adventures; others have fallen out of favour, but remain fun techniques that can bring history to life in a dazzling way.

## IRON PYRITES

The ancestor of more modern sparking devices, the wonder element of iron pyrites blessed our early forebears with the ability to make fire without the greater physical effort required by friction techniques. When struck with a piece of glassy rock such as flint or even another piece of pyrites, it produces dull red sparks that are effective in lighting certain tinders.

In Greek, the word 'pyrites' comes from 'pyrites lithos' which literally means 'stone that strikes fire'. Today many people know it by the nickname 'fool's gold' which was coined by early gold prospectors. It can be found occurring naturally, particularly in areas of limestone, shale and coal. In Britain, it is not difficult to find and there are certain beaches that are particularly good places to search for nuggets. It is not found every-where though, so in the past it was no doubt traded for its important fire-making ability. The best sparks are struck from the exposed internal material. Having acquired a nodule, start by splitting it open with a cold chisel to reveal the beautiful starburst formation inside. Sparks can be struck from this face. Alternatively, sparks can be struck from the exterior once you have abraded through the oxide layer. The most effective striking edge that produces the greatest quantity of sparks is a blunt and rounded one. A sharp edge will also work but the sparks produced are limited in comparison, and this method is therefore more time-consuming. The aim should be to maximise the area of contact between the two materials

OPPOSITE Set of materials for making fire from sparks, Batek tribe, Pahang state, Malaysia.
TOP An old cloth used to contain the fire set.
BOTTOM A selection of stones and a striker made from a scrap of carbon steel.

during a strike. Bear in mind the sparks produced are not very bright and can be difficult to see in daylight. If you want to experiment to see the quality and quantity of sparks you are producing, do so in a dimly lit room.

## HOW TO MAKE FIRE USING IRON PYRITES

To make fire, a spark needs to be caught in some tinder, and this needs to be very carefully prepared. Sparks from iron pyrites are much cooler in comparison to those produced by modern devices, and therefore the variety of tinders that can be used is greatly reduced. It is certainly not possible to ignite a bundle of dry grass or similar tinder with such sparks. In Europe, our ancestors would most commonly have used a piece of the trama layer from a horse hoof fungus (as described in Chapter 1).

I    To set a piece of horse hoof fungus smouldering, a marble-sized mass of extra-fine fluffy fibres needs to be scraped up from its surface using a sharp edge.

II    Now, holding the pyrites nodule steady in one hand directly above the tinder, bring the striker down, aiming for a glancing blow with a committed action. Sparks should fall delicately down from the impact and land on the tinder. Do not be disheartened if it doesn't catch immediately; keep striking and watch the tinder carefully. One spark will inevitably land a direct hit and will start the tinder smouldering.

III    This should then be encouraged, with a few gentle breaths, to grow into a substantial ember before being transferred to a fibrous tinder bundle and blown into flames.

Other tinders can also be used in the same way. These include cramp ball fungi, chaga, char cloth and even very dry punk wood. In fact, punk wood is very easy to ignite because a broad piece can be collected, which means accuracy is less important – wherever sparks land, there is a good chance it will catch and smoulder.

In the past, people would have carried a small leather bag containing a piece of iron pyrites, a flint (or similar) striker and some prepared tinder, which although it could have been collected from the bush was usually carried and replenished when needed – as we carry matches today. As described in Chapter 1, Ötzi the iceman was carrying just such a kit. The horse hoof fungus tinder he was carrying had particles of iron pyrite embedded amongst its fibres – remnants from previous fire-making sessions – but an actual nodule was not found with the rest of his fire kit, or near to the site where he fell.

Today, using iron pyrites to make fire is fun to try, but things have moved on. For travel in the bush, modern sparking devices are both easier to carry and superior in function.

**ICE AGE FIRE-MAKING KIT**
FROM TOP birch bark; horse hoof fungus; prepared horse-hoof trama tinder; iron pyrites; flint.

## STEEL

Once carbon steel could be produced, the iron pyrites nodules that formed the crucial part of everyday fire-starting kits began to be replaced by metal strikers. These were often beautifully made by the local blacksmith, and today such strikers are still available. The main improvement these new strikers offered was a hotter spark in greater quantities, making them quicker and easier to use. Other than that, these new strikers did not offer any major practical improvements. They were certainly more aesthetically pleasing than a crude lump of pyrites, though, and, in some places, easier to replace. The desire to upgrade personal belongings was no different back then.

In central Asia, most notably in Tibet and Mongolia, fire-making equipment was carried in a small leather pouch. This contained tinder, comprising downy fluff from catkins mixed with a small amount of powdered charcoal, and a sharp piece of stone riveted to a curved steel striker. These 'chuckmucks' were often intricately embellished and formed a neat little package that had everything needed to produce an ember. This steel strike-a-light technology did not really change at all until the development of the first reliable friction match in the early nineteenth century. Today you can purchase a striker, or for a fun project, make your own. In contrast to using iron pyrites, to make fire with this method you must instead pinch the tinder on top of the glassy rock and above a very sharp edge. When you strike this with the steel the sparks will be shaven off and will fly upwards onto the tinder. In an emergency, even the back of a carbon steel knife blade or a similar tool can be pressed into action.

**ABOVE** Modern European steel striker.
**OPPOSITE** Tibetan chuckmuck with flint and catkin/powdered charcoal tinder.

## HOW TO MAKE
## A TRADITIONAL STEEL FIRE STRIKER

All you need to make your own striker is an old file or some other scrap of high-carbon steel. Car-boot sales are good places to look if there is nothing in the shed.

   |    Firstly, heat your steel in the embers of a very hot hardwood fire until it glows cherry-red.

   ||    Very carefully remove it and let it cool down completely at the edge of the fire. This simple process will have changed the hardness of the steel; it will be much softer, making it easier for you to cut and shape it to the desired dimensions.

   |||    If it was an old file, you can use another file to remove the rough exterior and make it smooth and shiny. You can also cut it with a hacksaw if you wish to shorten it, and you can reshape the ends with a file. Be as creative as you like.

   ||||    Once you are happy with your striker's appearance, heat it in the fire once more until it glows cherry-red again.

   ₩    Remove it from the fire and quench it in a bucket of water preheated to 40°C.

   ₩ |    The exterior of your striker will be blackened with soot, so once it is completely cool, finish it with a good rubbing with emery paper. Your striker is now ready to use.

## FERROCERIUM

Of all the fire-lighting tools available to us for wilderness travel, ferro-cerium is the first choice bar none. It is a robust metal rod that when scraped with an edge harder than itself produces sparks of 1,650°C which cause a wide range of tinders to immediately burst into flame. Even dry wood that has been shaved fine enough will catch light. It takes a long time to wear down, so it is highly unlikely to run out during a trip; it has no small parts to malfunction or lose; and it doesn't matter if it accidentally sits in a pool of water all day – pat it dry on a trouser leg and spark away as normal. It is lightweight and compact, so you can even hang one around your neck on a strong nylon cord and stash several in different places in your equipment.

This alloy, which originally comprised 70 per cent cerium and 30 per cent iron, was patented in 1903 by Carl Auer von Welsbach. Shortly afterwards, manufacturing began, and this metal provided the essential 'flint' in cigarette lighters. By the mid-twentieth century, short, thick rods of the alloy began to be included as an important fire-making tool in military survival kits. Today, rods based on this amalgam are produced to cater for the needs of outdoors men and women as well as the military, and are a vast improvement on the earlier designs which were notoriously difficult to use. More robust rods are produced nowadays that easily create a shower of sparks when scraped. A word of warning: ensure that you buy a good-quality one from a reputable company – there are cheap copies available that either break or do not work. It is also important to point out that although ferrocerium rods will work even after a thorough soaking, if left wet constantly for many days, the metal will begin to crumble and it will become useless.

## HOW TO MAKE A SPARK WITH FERROCERIUM

Most rods will come with a scraper, but this will fail to produce the type of spark required to ignite some stubborn tinders that require one large, hot, lingering spark. To achieve this, you need to scrape quite a large piece of metal off the rod, and to do that you need to exert a lot of pressure. As always, the secret is employing a well-practised technique with the correct tool. You will achieve the best scraping action with the back of your knife, near the tip.

|    Start by placing the tinder on a firm, steady surface. If you are right-handed, hold the rod in your left hand, resting the tip right next to the tinder.

||    Holding your knife in your right hand, place the back of the blade onto the rod (do not use the blade edge). Use only the last few millimetres of the back of the knife, near the tip – the area where the bevel meets the back. Your right hand is responsible for pushing the knife against the rod. Your left thumb is solely responsible for exerting the pressure needed

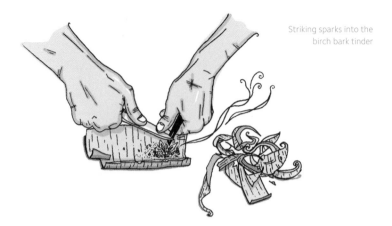

to push the knife along the rod to create one or two large sparks. If you practise doing it in this manner you will produce sparks that can light any of the tinders listed in the first chapter. Do not be drawn in by the impressive shower of sparks given off by any scraper that accompanies the rod when you purchase it – these will struggle to light anything but the easiest of tinders. There really is no substitute for the technique described above.

## BAMBOO

I came across a rather obscure method of producing sparks to make fire during my time spent in Waigeo, an island off the north-west coast of New Guinea. When lighters and matches are in short supply, the people there will often strike a broken shard of stone or crockery against the outside of a length of bamboo. During my visit, a local man described the technique to me, and it took several rounds of clarification before I grasped what he was saying. I thought there was perhaps some mix-up due to the slight language barrier; I could hardly conceal my amazement when he returned from the bamboo copse just behind his house and proceeded to produce sparks by striking a culm of dry bamboo with a piece of broken old plate. He told me that sparks could even be produced by striking bamboo on bamboo when no stone or porcelain could be procured.

During this journey, I was reading Alfred Russell Wallace's *Malay Archipelago*, which briefly mentions the indigenous people using this technique in Ternate, an island 400km to the west of Waigeo. I have also come

OPPOSITE Bamboo strike-a-light set.
Culm of bamboo, porcelain fragment and fishtail palm scurf tinder, Gam Island, West Papua Province, Indonesia.

across this technique in the Philippines, used by the native Batak people in central Palawan. Having conducted further research back in Britain, it appears there are various references to this technique occurring in many places other than those in which I have observed it: from northern Borneo eastwards to northern Sulawesi, across Maluku, and all the way to mainland New Guinea.

It appears that only certain species from the *Schizostachyum* genus of bamboo will produce sparks in this manner. The species I observed in Waigeo was growing near a saltwater lagoon and had a rough exterior similar to an extremely fine emery paper; the nodes were further apart than on most other bamboos.

For tinder, the dry scurf from a fishtail palm is used (as with the fire piston). Sometimes finely powdered charcoal is added to this to increase its combustibility. Traditionally, this tinder, along with the stone or porcelain striker, was carried in a bamboo box that was kept tethered to the length of striking bamboo on a string.

## HOW TO MAKE FIRE USING BAMBOO

I  Find a dry piece of the correct species of bamboo, at least 30cm long and about 2cm thick.

II  Hold a shard of porcelain or stone about 3cm in diameter, pinching a small amount of tinder to the top of it with your thumb, ensuring the tinder is close to the edge.

Striking sparks
from bamboo

III  Now strike the bamboo until a spark catches the tinder. Use either a quick glancing strike or a longer action that remains in contact with the bamboo, as you would when striking a match on its box. This should only take a couple of strikes but there is a knack to it, as usual, and it will take some practice to become proficient. The smouldering tinder can then be transferred into the fibres of a dry coconut husk and blown to flames.

OPPOSITE Batak house, Palawan Province, Philippines.

# CHAPTER 9

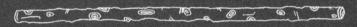

# SUN, ELECTRICITY
## & CHEMICALS

**HAVE YOU EVER THOUGHT ABOUT THE RISK OF STARTING A FIRE IN** your home just from the power of the sun? It's not an urban myth; light refraction is a real cause of fire that can easily spread out of control. From fish bowls to glass doorknobs and jam jars, the energy from the focused rays of the sun can be extremely dangerous, and if directed towards objects such as curtains, clothing, paper and furniture, can have potentially disastrous consequences. In 2015, sun rays shining through an empty glass Nutella jar filled with a child's Loom bands sparked a huge house fire in south London, destroying a home. Between then and early 2017, London's fire brigade stated that 125 fires had been triggered this way – and they have warned that the risk exists in winter, as well as summer. A constant flow of thermal energy from the sun's rays bathes every square metre, and whilst this is too weak to ignite certain materials, if the rays are concentrated, the energy becomes intense enough to surpass the threshold for combustion.

While deadly in an unexpected situation, harnessing fire from the sun's rays allows you to make fire without an ignition source, and can be a key survival skill when out in the bush. As a child, I soon realised that my magnifying glass came in very handy in my backpack when I was out and about. As well as using it to study insects and plants, it could be used to burn things and spark an ember to make a fire. It was comforting to know that I could use it over and over again with the knowledge that it would never run out.

I learnt more about using the power of the sun's rays to make fire during my first trip to Namibia. The time spent with the bushmen (described in Chapter 2) was at the very end of the trip, but earlier in that same excursion, we had focused on bush safety, emergency desert survival and tracking skills with Ray and Bob. Bob had many stories of his time growing up in Uganda and Kenya, and it was a pleasure to listen to and learn from him. After a couple of days on the road, criss-crossing the rugged landscape, we arrived at Hobatere – a 9,000-hectare reserve home to an array of African species, including elephants, lions, cheetahs, giraffes, eland and Hartmann's mountain zebra.

Our campsite was discreetly nestled in the landscape and was very private. We were surrounded by 'kopjes' – rocky, bouldery outcrops that occur in some areas in the bush, which are a common look-out and daytime sheltering spot for leopards and other creatures. The plan was to stay for five nights to learn the essential, everyday skills of travelling in this environment, and also emergency skills for surviving a crisis in the bush. I remember setting my tent up with some sense of trepidation, recalling that we hadn't crossed a fence or similar boundary since spotting a huge male lion out of the window of the 4×4 on the way there. He had a very dark, almost black mane and paused to gaze over at us as we crawled and bounced along the rough track. He glared as if looking into our souls – his mouth slightly open for a few seconds – before continuing his evening amble and blending into the drab hue of the thorny thicket.

I rolled out my mattress and puffed up my pillow, then found my head torch and hung it around my neck for later. The sun was going down and I knew that darkness would soon fall, and not just any darkness but the pitch blackness of the African bush. Having zipped up my tent door, I moved towards the voices I could hear. We all gathered around the fire waiting for dinner. Our Kalahari bushmen guides were amazing cooks. They seemed to be able to make the most delicious dishes from apparently very limited supplies, and all cooked over a campfire. Local game such as oryx, ostrich and kudu were regularly dished up.

Bob and Ray talked about encountering dangerous animals in the bush. Some species require different tactics from others and it is important to have a knowledge of these things if you plan on spending time out on foot. Binoculars are important because they enable you to scan the bush ahead to check the shadows and hard-to-see areas for a resting pride of lions, or a buffalo or hippo, for example. A means of making fire, a water bottle and a knife were other essentials that we had to have with us. If we were armed with this gear and got lost, or otherwise caught out for a night or two, we would be able to light a fire to provide warmth and light to combat the freezing nights (there was frost on my tent most mornings), to offer some sense of home and to help keep dangerous animals at bay.

We would have water to keep us hydrated and a knife for all the usual jobs, but also as a last resort should we be attacked by a leopard.

We had to light a fire using the rays from the sun, show that we could find water in the bush and signal for help using a mirror. Ray produced a small metallic object from his pocket – a parabolic mirror. He had collected some dried elephant dung which is extremely good tinder, and proceeded to put a pinch of it onto the device. As he pointed it towards the sun, I could see a small, bright, white spot of light on the backside of the tinder, which immediately began to give off a blue wisp of smoke. It was swiftly transferred to a bundle of dry grass and blown into flames. He told us it was our turn, and it wasn't long before people were returning to camp carrying mounds of fibrous elephant dung and bundles of dry grass. Ray came over to check my material selection.

Being able to confidently and quickly make fire is very important in this environment. One of the many reasons to light a fire in camp is to keep unwanted animals away. One night, Ray and I had ventured away from camp to watch a nearby waterhole in the evening (a haven for animals). I was carrying a pair of binoculars and a water bottle, and Ray had a spotting scope – a small portable telescope – attached to a tripod. We had been told to always move in pairs when walking away from camp. Having spent an hour or so watching the various visitors come to drink, the sun set, and we knew that it was time to head back to the safety of our camp. Darkness falls much more quickly near the equator: it's as if God turns out the light. Amongst the shadows of the thorny scrub, there was still a hint of dark blue and pink just above the western horizon. Day faded into night so rapidly we could almost see it changing.

We walked along the game trail we had come along earlier. It was still just about light enough to pick out the route between the bushes and boulders. That is when I heard the most terrifying sound coming from the bush ahead of us. The deep, agonising pants and growls of a black-maned lion made us freeze in our tracks. We waited stock-still for him to make another territorial call so we could try to pinpoint where he was. My knees shook with nerves. I knew that lions had a completely different

OPPOSITE The vast, arid, thorny scrub of central Namibia.

character during the hours of darkness. If you encounter lions on foot during daylight, they will often move off before you see them, or will stand and stare for a moment before letting out a little growl, turning tail and slinking away into the landscape. They are not always the courageous fighters that you might imagine them to be – that title belongs to the leopard.

The lion bellowed again, this time louder – or was it closer? I couldn't tell. He had definitely moved because the sound came from a different angle. I felt the vibrations of his booming voice rattle through my body, and in between each roar I could clearly hear him panting and grunting. I felt the fear inside me begin to increase and I started to pace towards camp. I wanted to be out of there! To my amazement, Ray started fiddling with the tripod legs and before I knew it, was focusing the scope in the direction of the lion.

'I think we'd better head back,' he said, calmly but hastily. As if I needed persuading! 'He's watching us — move confidently and do not run.'

OK, I thought; our camp was only 400m or so away. I calculated I'd be there in five minutes. I could see the flickering glow of the campfire reflecting off the nearby boulders. It glinted tantalisingly near, yet far way, like a homing beacon. That short walk was like a nightmare: you want to run but feel as if you're stuck to your waist in treacle that slows you to a hopeless pace. Standing next to the glowing campfire as my pulse eventually slowed was the most welcome relief I had ever felt. The next morning, there were huge lion footprints in the sand, just a few yards from my tent. They dwarfed my extended hand as I hovered it over the footprints to try to gauge the size.

With a little resourcefulness and improvisation, the properties of the sun, electricity or chemicals can be called upon to produce fire quickly and with little physical effort. In a survival situation, they will probably be the first resort before a friction method is employed. Of course, outside of these scenarios they are also fun and educational to try, and are some of the most surprising techniques of all. As always though, they must be treated with respect and used responsibly.

## SUN

Even though the sun is 150 million km away from us, it can still start fires here on earth if its rays are manipulated by us, by means of a parabolic mirror, or a magnifying lens, before coming into contact with tinder. The drawback of these techniques is that they rely on direct sunlight, uninterrupted by thick haze or cloud, and that is not always available to us. Even so, it is another technique to have in the toolbox, and is well worth knowing about.

## MAGNIFYING LENS

Unless you are travelling in a remote area in order to study small aspects of the natural world, it is unlikely you will be carrying a purpose-made magnifying glass with you. However, most of the time you will be able to find an adequate lens incorporated into another item of equipment that can just as easily be used to produce fire. Here are some options.

COMPASS: This 'key to the wild' should never be far from the reach of those venturing into the bush, even in today's world of GPS and other electronic gadgetry. Some compasses come fitted with a small magnifying glass which is not only extremely useful for observing minute details on a map but is also more than adequate for starting a fire.

BINOCULAR, TELESCOPE OR RIFLE SCOPE: Whatever optics you use must have a small exit pupil diameter – 4mm or less is good. Set the focus to the closest setting and hold it so the objective lens is nearest the sun. If you are employing optics where the magnification has a zoom capability, set it as high as possible. It is also worth noting that it is not necessary to dismantle optical equipment in order to achieve fire – use it as-is.

**READING GLASSES:** These work exactly like an ordinary magnifying glass, although some will work best when tilted at a slight angle to the sun instead of facing it directly. They also need to be held further away from the tinder than a similar-sized standard magnifying lens. Not all prescriptions will work, so you'll need to experiment with yours if you wear corrective lenses.

## IMPROVISING A LENS

It is very easy to improvise a lens capable of making fire using everyday materials and water. If one of the lenses listed above is unavailable, it may be that you can resort to one of the substitutes below. The key to using them is to ensure that there is no air bubble or wrinkly surface interfering with the light as it passes through.

**BOTTLE:** For this you need a transparent plastic or glass bottle, or a similar container with a pronounced rounded section. The shoulder area of a standard 2-litre soft drinks bottle works extremely well, as do bottles with a similar feature – even small designs. Once filled with water and fitted with its lid, the bottle becomes a lens and can be used as such. Simply let the sunlight pass all the way through the spherical area and onto the tinder.

**PLASTIC BAG:** Again, for this to work, a transparent bag is needed. Fill it with water, then twist the opening at the top in order to squeeze the liquid into one of the corners. Keep twisting it until the bag bloats into a rounded shape and any wrinkles or creases disappear. You will be able to use this as-is, but be careful not to drip water onto your tinder.

**CLING FILM:** Drape the film loosely over a bowl or other small receptacle and allow it to line the inside. Gently pour water into it, gather up the edges and seal it by twisting. The rest of the process is the same as described for a plastic bag.

**LIGHT BULB:** you will need a standard household bulb – the rounder the better. Carefully make a hole through the metal cap and remove the filament. Pour some water inside, shake it and pour it out. This should clean the bulb of any internal coating that may hinder its performance. Refill the bulb completely with clear water, seal the end with a rag or something similar, and it is ready to use.

# HOW TO MAKE FIRE
## WITH A MAGNIFYING LENS

Unless you are focusing the sun's rays onto ammunition propellant or similarly highly flammable material, you will not produce flames straight away. The sun's focused rays will only cause tinder to smoulder. This ember can be blown into flames, either on its own, if you are using fibrous tinder such as a coconut husk, or once it has been added to a larger, fibrous tinder bundle. The tinder onto which you will focus the sun's rays should be as dark in colour as possible, because darker materials absorb heat and light much better.

### POSSIBLE TINDERS

Dark brown, gnarly tree bark scraped into a fine powder and formed into a marble-sized pile; dry, fibrous herbivore dung; crushed dead leaves; cramp ball or other dark-coloured bracket fungi; dark punk wood; coconut-husk fibres; old charcoal or the charred end of a log; char cloth; dark print on newspaper or cardboard; safety match heads (if the striking surface on the box has worn out).

| Place the tinder onto a steady, dry surface that is in direct sunlight. If you are using fibrous tinder you must ensure it is tightly packed so as not to allow light to filter through its structure. A flat surface ensures the light will be focused on one area rather than being dispersed.

|| Bring the lens between the tinder and the sun so the light shines through it and projects brightly onto the tinder.

||| The secret is to move the lens to establish the 'sweet spot' – the point where you achieve the brightest, sharpest spot of intense light. It is like manually focusing a camera, but instead of turning a focusing ring, you are moving the lens backwards and forwards until you find the right place. With purpose-made magnifying glasses you will find this spot of light will be circular and easy to establish. With improvised lenses, however, it will sometimes be a streak of light and more difficult to identify the right point. A streak will still work, but it requires a little more patience. Once the sweet spot has been found though, it will work just as fast.

||||    Once you have found the sweet spot, hold it still so the light is working on one area of the tinder. If you have chosen suitable tinder it should begin to smoke in a few seconds.

||||    When it does, gently blow on the tinder to encourage the smouldering area to grow. If you are using a fibrous tinder, continue to blow on it until it bursts into flames. If you are using another tinder such as herbivore dung or a bracket fungus, you will need to transfer it to a fibrous tinder bundle before blowing it into flames.

Igniting some coconut husk tinder with solar energy
focused through a credit-card-type plastic reading lens.

## PARABOLIC MIRROR

A very effective alternative to a lens is a parabolic mirror. Simple and compact, they are available on the market as an emergency survival tool, their sole purpose being to make fire. The best design to choose is one with a tough housing that comes fitted with a lid, as this not only forms a place for emergency tinder to be stored, but also, more importantly, affords the mirror protection from scratches and the like that could reduce its effectiveness. These mirrors work by gathering light and reflecting it back into one small focal point; it is at this point that the tinder is held.

### IMPROVISING A PARABOLIC MIRROR

Although you are more likely to be able to improvise a magnifying lens, there are several ways to improvise a parabolic mirror, so it may be an option in a crisis.

DRINKS CAN: The underside of most drinks cans provides a concave surface that works very well, but they must be polished to a good shine first. You can achieve this by swirling a small amount of toothpaste or clay around with your fingertip before washing it off. This takes several minutes of work but does the job admirably.

THE REFLECTOR BEHIND A VEHICLE HEADLIGHT OR TORCH BULB: On some older designs this is possible, but modern lamp set-ups, both in torches and vehicle headlights, are becoming increasingly elaborate, which makes it both difficult to find the right-shaped reflector and trickier to remove it.

## HOW TO MAKE FIRE
## USING A PARABOLIC MIRROR

Purpose-made designs comprise a concave mirror with either a foldable or detachable tinder arm that projects out from the centre of the dish.

### SUITABLE TINDERS

Mostly the same as for use with a magnifying lens (see page 159), although due to the set-up of this device, the tinder must be solid, ruling out powdery or scraped tinders.

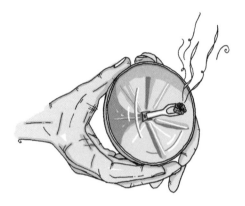

| When using an improvised mirror, you will also have to improvise a tinder arm. Using your fingers to hold the tinder in front of the dish may seem convenient, but it will cast too much shadow. Instead, skewer the tinder onto the end of a thin stick. As when using a lens, you will have to move the tinder back and forth until you find the all-important sweet spot.

|| Place a small amount of tinder on the very end of the arm, ensuring it does not block too much light from reaching the dish.

||| Aim the dish at the sun. When you see a small spot of intense light projected onto the underside of the tinder, hold the device steady in that position.

|||| After a few seconds the tinder should start to smoke, and from then on the process is the same as when using a lens.

## ELECTRICITY

In order to produce fire from electricity it is not necessary to have access to a large current. In fact, it is quite surprising what can be achieved with the small, everyday batteries found around you or in your expedition equipment. Bear in mind that using batteries in the following ways will reduce their lifespan and may damage them. Also be mindful that batteries can get hot when used in these ways.

Common small batteries such as AAA, AA, 9-volt and those found in mobile phones can all be used to produce fire; all you need to do is find a suitable material to conduct a current.

**FOIL WRAPPER:** For best results, you will need a piece of foil that is backed with paper. The wrapping from inside a cigarette box or a chewing-gum wrapper is ideal. This method works on the same principle as a fuse. Cut out a 7cm × 1cm strip and trim it in such a way that it tapers to 2mm wide in the middle. Use this strip to connect the terminals of a battery. In only a few seconds, the strip should snap at the narrow middle and a flame should appear. Quickly transfer the flame to some pre-prepared tinder before the paper burns out. This can also be achieved with regular foil or thin wire; in this case, you must manually touch it to some tinder as you connect the terminals. Unlike using wire wool, this can be achieved easily with one AA or AAA battery on its own.

**WIRE WOOL:** If you face a crisis on a vehicular expedition, it is worth looking in the repair kit or toolbox for some wire wool. A fine grade is the most effective. Take a piece about the size of a golf ball and tease it out a little so it is no longer tightly packed. Touch it with both the terminals from a battery and it will start to glow orange, spreading quickly. Once you see this, remove the battery and place the smouldering wool in contact with some fine tinder before blowing it to flames. If you only have AA or AAA batteries, one on its own will not make the wool glow – you will need to join two together.

**VEHICLE BATTERIES:** Larger batteries such as those found in vehicles are capable of producing sparks that are more than adequate to ignite any of the spark tinders. Disconnect the battery and connect jump leads to the terminals, then bring the other ends of the leads near to each other. Just before the ends touch each other, you will see a spark as the current jumps across. If you apply this near to any of the spark tinders, they will ignite.

Potassium permanganate and sugar mixture ignited by pressure from a stick.

# CHEMICALS

There are many combinations of chemicals that produce enough heat to make fire, but only two that are likely to be options for the wilderness traveller. The key ingredient in both of these reactions is potassium permanganate, a chemical that can sometimes be found in first-aid kits and has long been used as a disinfectant for wounds, fungal infections and even salad ingredients. Both reactions produce a very hot but short-lived flame, so you must be ready to catch it with some tinder.

## POTASSIUM PERMANGANATE & SUGAR

|     Put a teaspoonful of potassium permanganate and a teaspoonful of sugar onto a dry surface, and mix the two together thoroughly.

||     To ignite it, quickly draw a smooth, dry pebble or blunt-ended stick across the powder whilst simultaneously pressing down firmly. The aim of this action is to crush some of the crystals together and drag them across the surface. Imagine you're striking a giant match. You may need to do this a few times, but when you get it right the pile will begin to bubble with orange sparks before erupting into flames. Add your pre-prepared tinder quickly.

## POTASSIUM PERMANGANATE & ANTIFREEZE

This reaction is slightly different because the combination will ignite spontaneously, without the application of friction.

|     Put a teaspoonful of potassium permanganate crystals onto a dry surface near to where you wish your fire to be.

||     Drop half a teaspoon of a glycol-based antifreeze on top of the crystals and mix it a little with a twig. Hold your tinder and be ready to place it gently on top once you see flames. The speed of the reaction tends to relate to the weather conditions. If it is a warm, dry day you will get a fast reaction – perhaps ten seconds or so. On a cold, damp day it will take longer – sometimes a minute or two. However, once you have mixed the two together, always assume it is going to burst into flames imminently and be alert. It is a sudden and fierce reaction.

# MATCHES

Of course, one of the commonest ways to make fire in the modern world – the match – relies on chemicals too. Even though the availability of more convenient, inexpensive and reliable lighters has flooded the market in recent decades, matches continue to be used in great numbers. In fact, well over a billion matches are struck around the world each day. The first self-igniting friction match was invented in 1826 by British chemist John Walker. Prior to that, there were many designs of matches, but they were inconvenient, and often both dangerous and unpredictable to use. His invention continued to be improved upon during the remainder of the century.

In a wilderness setting, matches are a more reliable fire-lighting tool than lighters due to their simplicity. This is not to say that matches are perfect and lighters have no use whatsoever. I carry a lighter with me on any trip into remote country because they are convenient and last longer than an average box of matches. However, they do have moving parts and are more likely to go wrong. There are several types of match available:

STRIKE-ANYWHERE MATCHES: Although safety concerns have reduced the everyday use of this type, in the bush they are still the best option. Their advantage is they can be ignited by being struck on any rough surface, unlike other types which rely on a special chemical surface.

SAFETY MATCHES: These can only be ignited if struck on a special striking surface. This takes away the potential problem that strike-anywhere matches can suffer from – rubbing together and igniting in the box. The problem with this type is that the matches themselves are almost useless without this striking surface as it is not possible to improvise a striking surface in the bush. Also, the striking surface is prone to damage; the very action of striking a match erodes the surface, and on several occasions I have been unable to ignite a match even though I had only used half the box. Also, if this strip gets damp it will turn to mush. It is worth remembering that if the striking surface is ruined, safety matches can still be ignited with direct heat of some form (for example, by using one of the solar methods described earlier).

LIFEBOAT MATCHES: These have a waterproof coating on them as well as a much longer head than regular matches and as a result, burn fiercely for several seconds. This results in a flame

OPPOSITE Mount Tavurvur, East New Britain Province, Papua New Guinea.

166

that is highly resistant to strong wind and rain. They are a good option if you need to regularly light fires in very windy conditions, but are not a type I ever use. Even in extremely windy conditions it is possible to ignite tinder with regular matches, if the correct technique is applied. They are also bulkier, and are overkill for all but the very worst of conditions.

**PAPER MATCHES** This type of safety match is made of paper and comes attached inside a folding booklet. Their paper stems can very easily become damp, and they are rather fiddly to use, especially with cold hands. They are not suitable for serious adventures into wild country.

**STRIKEABLE FIRELIGHTERS** These are actually firelighters made from wood composite complete with a strikeable match head. They are impractical to carry in large numbers for everyday use in the bush but are a good emergency means of making fire – as are most firelighters. You may wish to tuck two or three of them into your outfit before departure, where they can be called upon to hasten the fire-lighting process if needed.

### HOW TO DRY MATCHES

If your matches seem to crumble when you try to strike them, they have probably got slightly damp. If they are not completely soaked, you can dry them out by running them through your hair a few times. You can sometimes get slightly damp matches to ignite by changing the way you strike them. Try stabbing them more directly onto the striking surface rather than drawing them along it.

If your matches have become completely saturated but have not turned to paste, you may be able to dry them out and salvage them. A good way to do this is to stick them onto a piece of adhesive tape, ensuring there is plenty of room in between them. This strip of tape can be hung up in the breeze or high up in the warm air of a cabin to dry. If you don't have tape, anchor them where the sun and breeze can get to them without blowing them away.

### HOW TO STRIKE A MATCH

When we are far from the reach of outside assistance, the consequences of even the smallest of mistakes can snowball into life-threatening situations. It might seem unnecessary to describe how to strike a match, but there is a right way and a wrong way to do so, and not knowing how has sadly cost lives in the past. We lose dexterity and feeling in our fingers when we get cold (we have all experienced the difficulty of doing up a button or retying a shoelace after a snowball fight). This means we cannot detect how much force we are applying and can easily break matches.

Most people strike matches by applying pressure with their thumb or index finger held at the middle of the stem. When your hands are cold, this will almost certainly lead to you snapping match after match without ignition. With a loss of dexterity, it may be impossible to pick up and hold a broken match stem.

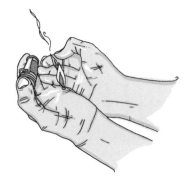

I    The answer is to hold the match at the very end, between your thumb and index finger and support the head by placing your middle finger on top when you strike it. This way, the pressure runs along the stem instead of across it.

II    When the match ignites, immediately remove your middle finger, and keeping hold of the box, bring the match into your cupped hands to protect it from the elements. Aim to have the match cupped before the head has finished flaring.

III    Wait a few seconds until the stem has caught and the flame is strong before introducing it to your tinder.

IIII    Keep the match cupped until the very last moment. Sometimes in very strong wind, however hard you try to protect it, the flame keeps blowing out. In this situation, I sometimes strike a match and immediately introduce it to the tinder in one swift action – the aim being to have the match amongst the tinder while the head is still flaring.

IIII I    Another trick if you need extra flare is to hold two matches together as you strike them.

IIII II    Once your fire is underway, put the matchbox safely away. Never put spent matches back into the box; burn them on the fire.

**SOME OLD FRIENDS**
that have been with me on many a journey.

CHAPTER 10

# FROM FLAME
# TO FIRE

**FIRE IS A BASIC ELEMENT OF THE UNIVERSE. IT IS OFTEN CONSIDERED**
a friendly and comforting force, as anyone who has huddled around a
campfire in the middle of the night will tell you. Yet fire's energy can also
be dangerous, deadly and threatening, consuming everything in its path. It
plays a central role in many myths, folklore and legends around the world.

The thunderbird is a supernatural creature of gigantic propor-
tions and huge strength, depicted in various forms in North American
indigenous myths. The thunderbird is said to produce thunder by flap-
ping its wings, and lightning by opening and closing its eyes, bringing
fire to the land. It was a widespread belief in the ancient world that these
birds controlled the weather, creating ferocious storms and 'thunderbird
fire'. This fire was said to rejuvenate the land, attracting animals and new
growth. In Pikangikum, a Native American reserve in north-west Ontario,
Canada, the area has strange, random piles of rocks that are believed to be
thunderbird nests.

I remember helping Ray Mears to lead a group of twelve canoeists
there several years ago, alongside a well-known Canadian canoeist, Becky
Mason. We were heading out into the bush for a fortnight-long journey,
and had bumped and crawled along a bouldery track for at least a couple
of hours in a 4×4. At last, we reached the end of the trail and the vehicle
could go no further. We were now in a land where the canoe ruled over
all other means of transport. Having disembarked, we stretched our legs,
hastily unpacked the canoes along with our bags, and carried everything
down to the nearby lake shore. We had left the main town later than we
would have liked, but we still had enough daylight to paddle the kilometre

or so to an island we found suitable for camping. Once there, the plan was to find a good place to put our tents up, then get the kettle on and have some dinner.

Almost as soon as we had placed the last canoe on the beach, the dark clouds that had been looming on the horizon made their way over to us. The rain poured, and the thunderbird was among us. Dazzling white lightning flashed all around and thunder boomed above our heads, pealing off into the distance. There was no way we could safely get to our island in these conditions. We waited and planned. If there was a break in the storm, we would get the canoes and make a quick dash for our island. We considered turning our boats upside down and setting up camp where we were – but there were signs of black bear, including fresh droppings full of berries and claw slash marks on trees with the resin still running out of them. We knew it would be unwise to stay. With paddles in hand, we looked up at the sky in an attempt to predict its next move, watching, waiting for the slightest hint of a break.

Suddenly the sky brightened a little as the clouds parted. The rain which had been pounding down began to ease, and we took that as our cue to move. Everyone pushed their boats out into the water, jumped in and began to paddle steadily. Although the lightning had passed, its presence could still be felt. It was as if we were trying to run across a garden and up to the door without waking the guard dog. We looked like a war party as we dug our paddles deep into the dark water and pulled back hard on them. As we steered our boats towards the island, I looked over my shoulder and saw our party close by, all keen to make it ashore and relax.

Midway on our short journey, with our party hugging the shoreline as closely as possible, the clouds gathered again and stole the last of the daylight from us. The rain's intensity increased; my boat had a good inch of water in the bottom, swishing around my knees and wicking its way up my trousers. A bolt of lightning struck out randomly and hit the trees in the distance. A few seconds later, the clap of thunder we had all been anticipating served as the crack of a giant whip that made our paddling twice as fast, along with our heart rates. The storm had returned and we all sensed the danger. Immediately, Ray and I turned our boats to point at the nearest refuge and led our group to a small island in the middle of the lake. It was the safest thing to do. Being only a hundred metres away and paddling fast, we reached the shore rapidly. There was no gentle beach, just a rocky edge laced with a spiky tangle of fallen spruce trees. We dragged our boats up as far as we could and tied their painters securely to the nearest hold. The gear was unpacked and a chain of drenched silhouettes passed it all along one by one. I was blinded by waving torchlights which illuminated the rain in their beams, enhancing its ferocity.

I shouted to the group to find a place to get their tents up, and they set to. A flash of lightning lit up the whole island, and then another shortly after. The island was barely 70m across, and was completely covered by a tangle of fallen trees that had been blown over in powerful winds the previous winter. Everywhere my torchlight fell, I saw nothing but soaking wet trees lying horizontally at knee height. It seemed hopeless to find a place for a fire, let alone a tent. Everything was soaked except our sleeping kit; the forest, our bags, our clothes. But we knew without a doubt that lighting a fire was the most important thing to do. We needed to bring the group together and provide a sense of home in the appalling circumstances in which we found ourselves. But how? To most it would have seemed as if there was not a single piece of dry firewood left in the forest. But we could easily make a flame – we had dry matches and lighters in our pockets. The most important thing was to transform that tiny flame into a self-sustaining fire.

Ray cleared some branches away from a flat spot and strung up a tarpaulin between a few trees that had remained standing while I sawed down a thin, dead, standing pine tree. I dragged it back over to the tarp, sawed it up into sections free from knots and split them into thin splints with my axe. Working together as the rain rattled on the thin roof above our heads, we shaved the perfectly dry wood from the centre of the sticks and created a mass of tight curls on half a dozen of them. A platform of split wood was set down on the wet, moss-covered ground and these 'feather sticks' were arranged carefully on top. Ray produced a handful of dry birch bark and placed it on the platform. Not wanting to risk getting my matches wet, I scraped sparks from my trusty ferrocerium rod onto the birch bark and pushed it beneath the pile once the delicate flame had taken hold. We watched it grow rapidly, and once it was ablaze we added extra dry splints on top.

Soon, the tents were up and the kettle was dancing over the flames; at last we could relax in the knowledge that everyone was safe and dinner wasn't long away. We had established camp by one simple act, and in doing so had transformed a dank and uninviting place into a relatively comfortable home for the night. The rest of the group, some of whom were experienced in the bush, knew instinctively how to set up camp. In the middle of the Canadian wilds our glowing fires, radiating their heat and cheerful light, must have made the thunderbird look twice. I was still feeling pleased with myself when I opened my bag to search for a couple of hot chocolate sachets and noticed an equipment pouch half-full of water. I emptied it out and, to the amusement of my companions, discovered my passport, now a saturated sponge. It was my own fault; I sat out until the early hours, drying page after page next to the flames.

# THE KEY TO MAKING A FIRE

It is one thing being swift to pull off one of the marvellous techniques described so far and produce a flame; it is another altogether to secure and extend its blessing. That feeble ambassador of fire shows its face for the briefest of moments before vanishing, and returning both the surroundings and human morale to a dismal state. More than any other stage, this is where those that really understand nature's gift of fire are separated from the rest: not because they can start a fire, but because they know how to make a fire last for certain.

Of course, like all things, once you have delved into it, there ceases to be any mystery and it can be learned by anyone. Yet it is an art: a display of subtleties and finesse that goes unnoticed by the novice. Most fires kindled by the inexperienced have passed very close to failure, and are often only successful thanks to good weather conditions and lucky material selection, rather than a deliberate chain of decisions. For those learning, there will be more failures than successes – probability takes care of that – but novices will not always notice that it is they that are doing something wrong. It seems as if the decision is left to fate: will the fire burn or will it go out? Tuition on its own is not the answer; practice is, once again, the only real way. The best fire-makers and managers are those with inquisitive minds – the people who ask a question, and then ask another question because they are not satisfied with the answer.

Certain characters can tend a campfire better than others, regardless of experience: those that sit and fiddle with the embers or, while deep in conversation, brighten up the fire by nudging a burning log a centimetre away from the next with the toe of their boot to liberate a trapped flame. They tend to be good estimators and able to gauge things accurately by eye. Remember, while training and having fun at home, you can afford to make mistakes, but out in the wilderness, far from safety, the consequences of such errors can be great.

OPPOSITE The fleeting moment when fire takes hold of the kindling.

## WHERE TO SITE A FIRE

The first consideration once you have decided to light a fire is where to site it. Both safety and practicality are served in making good decisions on this point. If we are to light a fire safely and leave little trace of our passing, it is very important that we avoid making a fire on top of peat or decaying vegetation, because the fire will burn into the ground and spread. At best, this will leave an unsightly scar on the landscape that is difficult to clear up; at worst, it may allow the fire to travel underground, springing up on the surface at a later time and causing an uncontrolled blaze, potentially putting other people and the landscape itself at risk.

Always try to light a fire on mineral-based ground: soil, sand, rock or an exposed area of gravelly river bed. If there is any vegetation, clear a space back to the bare ground and make it big enough to ensure the fire cannot spread.

In addition to this, it is easier if we site a fire – and therefore our camp – near to any resources we may need. The necessities in any camp are usually going to be an adequate supply of firewood to accomplish the needs of the camp, and a water source, both for consumption as well as clearing away a hot fire site. If you are able to light a fire on an exposed piece of stony river bed when the water level is low, this will mean your fire site will be completely cleaned away once levels rise again.

In very remote wilderness, some of the fires we light will be in new areas where a fire has perhaps never been lit before. In some places – national parks and the like – there may be existing fireplaces which have been used by other visitors.

## HOW TO LIGHT A FIRE

The flames given off by a single flame or a handful of burning tinder are relatively weak and short-lived. Think of the lighting of a fire like the take-off of an aircraft; it is one of the most delicate and vulnerable moments either can encounter, but a necessary part of the process if a steady and certain outcome is to be secured. Just as the skill and experience of the pilot can ensure this, so too can that of anyone lighting a fire.

For a fire to grow, the flame must spread from one piece of fuel to the next, and this is where many problems can arise. As a general rule, never try to add any fuel to a fire that is more than three times the diameter of the current fuel's diameter. Unless the fire is very large, this will result in the fire losing momentum or even going out; the smaller fuel will burn out before it has transferred its flame to the new fuel. It takes time and practice truly to learn the capabilities of fire and understand its behavioural quirks and nuances. Follow the steps ahead, though, and you will be able to adapt the initial flame to your needs with certainty, even if you have never lit a fire before; you will be well on the way to discovering the rest for yourself.

Remember, the key to success is proper preparation. If you are inexperienced and try to cut corners to save work, you are likely to fail.

START BY MAKING A PLATFORM

Wherever and whenever you light a fire outdoors, always lay a platform of dead, dry wood down on the ground. The fire will be lit on top of this and will destroy it in a few minutes, so it doesn't need to be anything elaborate: a dozen or so 30cm-long sticks of finger thickness, or splints split out from a log and laid next to each other, is all that is necessary. There are three main reasons for this platform. Firstly, it separates the early stages of the fire from the ground which may be cold and damp – fire's greatest enemies. Secondly, it allows air to flow underneath the fire, which will result in a quicker and more complete combustion. Finally, once the fire has been underway for a couple of minutes, the platform itself will begin to burn, which will provide a good hot heart to the blaze. These things can make all the difference in tough conditions. Of course, if you are in a warm, dry place and the ground is not damp, you can get away without making a platform, but it takes hardly any effort and is a good habit to get into.

Contrary to the popular image of a campfire, there is no need to surround the fire with stones. This blackens the stones and leaves a lasting impression of your passing upon the landscape. This is inconsiderate camping, so unless you are using an existing and established fireplace surrounded by rocks, you should not create one.

## LIGHTING A FIRE WITH THIN TWIGS

Unless you are somewhere that receives a great deal of rainfall, most of the time the easiest and most common way to get a fire going is with thin, dead twigs. When travelling in the forest, there is an abundance of wood in different stages of life and decomposition, and so there are always thin twigs available. The only exception is the jungle, where wood rots fast and it can be tricky to find suitable twigs.

The type to look for are the thinnest you can find – ideally no thicker than a match. If you look around in the trees you will see lots of old, dry, fallen twigs that have become caught up in the branches on their way down to the ground. These are ideal because having been kept off the ground and aired, they will be the driest. Other good places to search are amongst the lowest branches on coniferous trees – these are often the thinnest you will find, and in dry weather can even be lit directly from a match, allowing you to conserve tinder.

| Choose twigs that are brittle and snap cleanly, and avoid living branches or those that have only recently died. Do not collect twigs directly from the ground unless there have been several warm, sunny days in a row.

|| Keep the twigs at least 40cm long and collect a bundle thick enough that you can only just get both hands around them.

||| Split this bundle in half and place one on top of the other in an 'x' shape directly onto the platform.

||||    Nestle some tinder right next to the point where the two bundles cross (see ABOVE), and ignite it.

||||     Wait until the twigs themselves are burning strongly and there are flames licking through before adding a similar-sized bundle of pencil-thick sticks.

|||| |    Once the flames have taken hold of those, add a bundle of thumb-thick sticks.

|||| ||    Once those are burning, you can continue this gradual stepping-up of the size of the fuel if needed, and manipulate the fire to your needs.

If there has been some rain and the forest is wet, this does not mean that this method is out of the question; in fact, wet twigs will still burn very well providing they are not saturated to the core. The bundles of twigs you collect need a couple of firm shakes to get rid of excess water, and you will need to use more tinder than usual as the flames will need to dry out the fuel before it catches.

## LIGHTING A FIRE WITH SPLIT WOOD

When birch bark is available in abundance and you have an axe with you, there is an alternative to thin twigs as a means of starting a fire.

| Split out a load of finger-thick splints from a dry log or, if you can find one, an old wind-blown pine stump infused with resin.

|| Place two splints in a 'v' shape on top of the platform with the opening towards you.

||| Nestle a good handful of birch bark peelings in the centre of the 'v' and ignite it.

|||| Now carefully rest other splints over the top, leaving a little gap between each one and the next to allow the flames to come through.

卌 Build this pile up in layers; each layer placed at 90 degrees to the previous. Although this method works outside just as well, it really comes into its own in a wood-burning stove where thin twigs are inconvenient in such a restricted area.

## LIGHTING A FIRE WITH SHAVED WOOD

In some circumstances, using the more usual thin, dry materials for starting a fire is not an option, either because they cannot be found or are wet. There are two situations when this problem may present itself. One is when you are travelling in the mountains, above the tree line or somewhere else devoid of trees, and the cabins you stop at only contain a pile of logs next to the stove with no kindling. The other is when you are travelling in the forest and there has been heavy and prolonged rainfall. Often the fastest and sometimes the only way that fire can be started in these scenarios is to manually make tinder and kindling material by shaving it from the dry interior of larger pieces of wood. This is one of the most misunderstood and misrepresented fire-making techniques there is, which is one of the reasons why it is also wrongly underrated. However, executed correctly, there are few other skills that can prevent unfavourable circumstances from deteriorating into life-threatening crises as well as this one.

There are a few ways of adapting the skill of shaving wood according to your circumstances, but whichever you choose, you need to start by collecting a standing, dry, dead branch or whole tree at least as thick as your arm and as long as you can find. Any wood not used in the actual process can be burned as fuel. Look for all the usual signs that tell you if the wood is dry or not, and select only the driest. If you are in a cabin you may have a pile of suitable wood already, in which case there is no need to collect anything.

### SHAVING CLUSTERS

The beauty of shaving clusters is that you do not need to split any wood. This means you can easily make them from a large diameter log, even if you only have a small knife. They are also more convenient when lighting a stove or other indoor fireplaces where space is restricted. Beginners will find these easier to produce than feather sticks because the shavings don't need to be as long.

|    Support the log so it is held vertically. This is the easiest position in which to carve it.

||    Take your knife and shave into the surface of the log repeatedly in the same place until you have made about half a dozen 10cm-long shavings that remain attached to it.

|||    Now, you need to remove the cluster of shavings from the log so they remain as one. Do this by pushing your knife deeply behind the last shaving and tapping the back of your knife with a baton of wood. Pile them up on your platform as you make them and when you think you have enough, add your next stage of fuel on top and set them alight.

Feather sticks are similar to shaving clusters, but you really need an axe or a similar tool in order to cut up and split the wood. Feather sticks have a larger mass of curls that are left attached to the stem.

    |      Cut the piece of wood into straight, clean, knot-free sections of about 45cm long. Split these down to create splints that are as thick as a thumb. These splints can be shaved with a sharp-edged tool, usually a knife, leaving the shavings attached at one end.

    ||      Support the splint of wood vertically by resting the far end on a log or firm ground and kneeling beside it.

    |||      Ensure your knife is sharp, then place it at the top end of the wood and run it down the length steadily, keeping your arm locked straight, to create a thin curl that remains attached at the bottom.

    ||||      Repeat this over and over until you have a ball of shavings attached to the bottom of the stick, or the stem becomes too flexible to support the pressure. If you are new to using a knife, you will in all likelihood cut off most of the shavings before they reach the bottom, but do not be deterred: these shavings will still be of use and you must go through this process in order to improve. The key here is knife control – so much so that a person's carving and woodwork ability can be accurately assessed from watching them carve a single feather stick. When they can carve a good one, they possess the skill to carve almost anything.

Safe, efficient method of making feather sticks

Don't carve too deeply. The aim is to finely shave long slivers that collect at one end, not gouge short, stubby protrusions all over the stick so you end up with something that looks like a Christmas tree.

Another way of carving them is to place your knife hand in front of your knee so the blade is horizontal and to then draw the splint of wood onto the blade with your free hand.

Igniting a feather stick to start the fire

|||| When you have half a dozen good feather sticks, arrange them in a 'v' formation, flat on a split-wood platform so all the shavings are together at the point of the 'v'. You can then set them ablaze by igniting a shaving right at the bottom of the pile before placing more split wood on top.

Note: In an emergency, feather sticks can be lit with a spark from a ferrocerium rod. To do this, scrape the stem of one feather stick to create a marble-sized mass of thin wood scrapings right up against the last shaved curl, then drop a spark into this. Avoid moving the stick until the flame is strong.

Sorting tinder, kindling and fuel carefully before lighting the fire will ensure that you have the right material at hand to add at each stage as the flames take hold.

## COMMON MISTAKES TO AVOID WHEN LIGHTING A FIRE

### UNSUITABLE FUEL

It is easy to make mistakes with fuel selection when you are inexperienced or working under stress in difficult conditions. Make absolutely certain that the wood you are collecting is dead and dry, and not merely dormant for the winter, or recently cut. Remember, when you saw firewood, the sawdust from the middle should float down and not be clumpy. As a rule, dry wood will feel light in weight relative to its size.

### BEING TOO SLOW

If you are too slow to add fuel, the fire may burn out before it has spread. There is a balance to strike here. You don't want to smother the fire and exclude oxygen by piling too much fuel on, but on the other hand there is no need to add twigs one by one. Handle it firmly.

## FIDDLING

Many healthy fires are loved to death. They are extinguished by too much unnecessary titivation, especially during the lighting process. Prepare everything as described, put a light to the tinder, and resist the urge to interfere with it. It wants to burn! If you move sticks around too much in the early stages, you will spread the heart of the fire too thinly and it will die. Keep the embers together.

## A FLAT FIRE

Flames like to burn upwards. If you place firewood on and make it too flat, it will smoulder and cause irritating smoke. Strike a match and hold it horizontally. Then hold it at a steep angle, and you will notice it burns faster and brighter. The same goes for a large fire: lay fuel on at a slight upwards angle.

## A TEPEE FIRE

The other extreme to a flat fire is this common misconception of how to build a fire: a tepee-shaped fire. It is often seen in illustrations, probably because it is easy to draw. In reality, it is neither easy to build nor reliable. Although it is a shape that burns brightly when it works, it is not a style I would recommend. The sticks are leant up against each other, and are therefore prevented from collapsing into the fire. This often results in the centre of the fire burning out while the sticks leant up on the outside are left merely scorched.

## PACKING FUEL TOO TIGHTLY

This can prevent a fire from burning efficiently, or can even smother it completely and put it out. When you lay fuel on, make sure the flames have air space to lick up between and around each piece. A good example of this is when you burn old newspapers or magazines in a bonfire. Even if they are thrown into a raging inferno that lasts for several hours, once the fire has gone out you can often rake through the ash and find perfectly legible, unscorched pages.

# WILD FIRES

## ABORIGINAL USE OF FIRE

Fire is an important symbol in Aboriginal culture. For thousands of years they have used it for warmth and cooking, as well as hunting and managing the landscape. Children are actively encouraged to play with fire from a very young age and set light to the bush, as doing this plays an important role in their lives – including making access through thick and prickly vegetation easier; encouraging new growth and the regeneration of plant life; attracting game for hunting; and for spiritual reasons. Aboriginal people are very aware of the right time of year and correct ways of using a fire. The frequency of burning and the time of year are chosen by the impact this approach has on food supplies and other survival factors. The overall effect of these fires is called 'fire-stick farming' and has created a light, regular mosaic pattern of burning, producing a varied number of neighbouring habitats at different stages of regeneration after fire.

Fire is a vital aid in hunting and used to direct animals, so they can be easily speared. Kangaroos, wallabies, emus and mala try to escape the flames, and sometimes a whole community works together to produce meat for everyone. Some will light fires while others wait in ambush downwind. Fire also acts as a signal. Aboriginal outstations are widely scattered over the vast landscapes, and Aboriginals travel thousands of miles from their camps to hunt and gather plants. The plumes of smoke help to identify who is working where, so it stops any one area from being hunted out.

Fire also has significant ceremonial and symbolic significance. In one ceremony, called a 'smoking ceremony', which is performed at important events such as births and deaths, indigenous Australians burn native plants to produce smoke that is believed to have healing and cleansing properties, and the ability to ward off bad spirits.

## FIRE ECOLOGY

When we see a wildfire, it is easy to think that everything is being destroyed, but some ecosystems depend on periodic fires. Bushfires caused by lightning are common in many parts of Australia, and in forested areas of the USA and Canada, but it is not always bad news for the land. A number of plants have evolved to find ways of surviving these phenomena. One of the most interesting adaptations is that some plant and tree species require fire for their seeds to grow. Plants like eucalyptus and banksia in Australia have developed what is known as a serotinous cone, covered in hard resin which only releases its seeds after the heat of the fire has melted the resin.

The same theory applies to lodgepole pine and jack pine cones in the Canadian boreal forests. Fire also produces favourable conditions for these seeds to germinate, preparing the seedbed and reducing competition from other plants. Both jack and lodgepole pine depend on fire to regenerate. Jack pine seeds have been known to still be viable after exposure to heat as high as 535°C. Some other species and plants require chemical signs from charred plant material and smoke to break seed dormancy. A shrub belonging to the buckthorn family called ceanothus responds to the heat of the fire, and will release seeds. Plants like iris stone and mule's ear, meanwhile, store much of their energy in their underground bulbs and roots, so when everything above ground is burned, they are first to react to the ash that is rich in nutrients.

## HARNESSING FIRE FOR ADVANTAGE

In Australia, at least two birds of prey – black kites and brown falcons – appear delighted when a bush fire tears through an area. They will swoop down after the flames have relented a little to snatch panicking prey, including large insects, animals and frogs. Since there is stiff competition for prey, some will pick up burning sticks, and fly to an unburned area to drop their deadly cargo, deliberately encouraging further destruction, so they have to fight less for their meals. Researchers believe these birds are a 'third force' capable of causing bush fires (the other two being lightning strikes and man).

Some black kites and brown falcons have even been observed using this same tactic to set light to bush land on both sides of a road, flushing all sorts of creatures away from the terror and onto the flameless surface, only to be picked off from the tarmac and devoured by the birds. Although the predators have never been caught on camera in the act, many witnesses, including numerous Aborigines, firefighters and bush rangers have described how birds can carry smouldering sticks at least 46m without the fire going out or the bird being singed. Currently more evidence is being gathered exploring this phenomenon.

**ABOVE** Osprey, New Ireland Province, Papua New Guinea.
**OPPOSITE** Fire-blackened banksia seedhead, Western Australia.

# CHAPTER 11

# FIRE SET-UPS & MANAGEMENT

**I AM AWARE THAT I TOO AM ALWAYS LEARNING. PART OF ANY JOURNEY** is experiencing unexpected setbacks and problems. It is how we deal with these that is of the utmost importance. Strength and knowledge comes from our ability to find solutions, pick ourselves up and persist, even if the odds are stacked against us. In the world of bushcraft skills, complacency is the enemy. I continue to acquire knowledge and skills, even in the unlikeliest of places, and delight in putting these into practice.

One of my single greatest joys is the ability to head out to the remotest place for an adventure. My friend and colleague Rob shares this delight, so when he suggested a trip to the forest near Tampere, Finland's second largest city, I was in. His idea was to hire skis and head out. Rob and I had been on many trips together, hiking in the Scottish highlands and skiing once before in the mountains of southern Norway. We had also both completed winter survival training in the high Arctic, so we knew how to take care of ourselves in extreme cold. I jumped at this opportunity to dust off my winter equipment and get back out into the fresh, biting air of the Scandinavian winter.

We packed our rucksacks with some basic winter kit: gloves, hats, woollen layers, windproofs, snow shovels, axes, knives, torches and cooking pans, amongst other bits. We wanted to travel as light as possible so we decided not to take a tent or tarp or a cooking stove. Instead we planned to sleep in 'lean-tos', a kind of open-fronted log cabin built on Finland's forest trails and national parks. We also carried Goretex bivvy bags, inflatable sleeping mattresses and winter sleeping bags. For cooking, we planned to make a fire – simple, easy and what we were most used to – so we weren't weighed down with yet more kit.

As we exited the terminal at Tampere, I put my rabbit-fur hat on and glanced up at a digital thermometer on the side of a building: minus 11°C. I could feel ice crystals forming in my nostrils as we waited on the frozen pavement for a bus to the hotel. Particles of ice swirled and danced like smoke on the dry road as vehicles went past.

The map we had brought with us was at 1:100,000 scale, which had less detail than we would have liked, so the following day, after buying some rations, Rob went on a map hunt. Nowhere had a hiking map of that area in the scale we wanted, unfortunately; our map was still reasonably clear, though, and we were confident we could use it, especially as we planned to follow a ski trail which would be signposted at major junctions.

After a beautiful drive through endless forest, we arrived at our destination. Clipping into our skis, we hit the trail. It was a great feeling to glide on the hard-packed snow on the roads, which sounded like squeaky polystyrene beneath our feet, and we both soon got our ski legs back. After about a kilometre or so, we turned off the road and headed into the dense forest. We could see a track in the deep powder snow ahead. We also noticed the trail was marked every so often with a spot of red spray paint at head height on trees, which was very useful because our map's scale made it difficult to interpret certain features. We were making good progress and were less than a kilometre from the lean-to shelter when the trail suddenly became unclear. It had been snowing heavily since we left town, and the tracks from previous skiers had either been buried completely or had turned off and gone another direction back down the trail.

We both stood by a marked tree but could not see the next sign. We took a bearing from the map and searched along that direction, but still couldn't locate the trail. We began searching other openings between trees that looked as if a trail might fit between them; we scraped the ice off as many trees as we could to try to reveal a red mark, but no luck. There was no feature we could take a bearing on and head for – no lake or stream for a long way. We marked our last-known point, then moved out from there in different directions, probing possible trails and looking for signs. Half an hour later, we still couldn't find anything. It was hugely frustrating because we knew we weren't far away. By now it was pitch black, the temperature was dropping fast, and we were hungry, tired and thirsty, fighting the urge to panic. In these situations, it is easy to press on and convince yourself that

it will all make sense just around the corner. But 'just around the corner' inevitably turns into 'just around the next corner', and much more than that, and eventually results in becoming totally lost – or far worse.

We put our head torches and warm coats on. Rob opened a flask of hot tea and I dug out a couple of Mars bars. Sitting on our rucksacks and sipping our drinks, we could see the ice crystals twinkling in the air like diamonds. The tips of my ears started to sting. It must have been minus 20°C, and there was almost a metre of snow on the ground. We knew the right thing to do was to 'STOP': Stop, Think, Observe and Plan'. Dehydration often plays a role in poor decision-making; very often if you stop, have a rest and take on some fluids, all becomes clear. Making the decision to stay put for the night can make everything more relaxed; the ground in which you stand becomes home in your mind. This is a very powerful feeling.

We decided we were going to pitch up. A fire was essential; without it we wouldn't be able to make water, cook our dinner or warm up. We had learnt during our survival training how to sleep out in the forest if pressed, but nonetheless, I felt apprehensive. We took our skis off and marked out the edges of a rectangular trench, large enough for us both to lie down in, with a space at one end for a fire. We started to shovel, sweeping the sugar-like snow out, piling it up on the sides in order to build a little wall up to add more protection from the wind. We dug right down to the moss-covered ground.

I waded through the snow a few metres and cut down a thick, dead, standing pine tree with my axe. We needed a decent amount of fuel and wanted some radiating warmth. Meanwhile, Rob collected some spruce branches and laid them neatly in the trench, which offered good, springy insulation from the ground. We sawed the tree into several 2m lengths and split the thinnest end into splints. We laid two of these long logs parallel to each other, filled the small gap in between with feather sticks and birch-bark peelings, and set them alight. Rob cut a sturdy, thin, forked branch from a nearby birch and pushed it into the snow near to the fire. He filled one of our steel cooking pans with snow and hung it on the stick, just above the flames. Soon we would have water and could hydrate our rations. As we were travelling without a tent or similar shelter we improvised a little roof over the head end of our trench. We pushed our skis and ski poles into the snow on the edge so that they overhung slightly and crossed over each other.

We draped our jackets over this framework; our roof was complete.

By now our long log fire was going nicely. We inflated our mattresses and slid them along with our sleeping bags into our bivvy bags. They were laid out on top of the spruce boughs. Finally, we could sit down and relax a little, regularly topping up the pan with snow. With the fire kicking out some serious heat, we stripped off some layers, dried out our gloves and saved the batteries in our torches.

Soon the pan was full up with water, which bubbled a little over the side, hissing as it landed in the embers. We poured some into our ration packets, made a couple of hot chocolates and had enough to fill one bottle up. We refilled the pan with more snow and as we waited for our dinner to cook, we studied the map and discussed what to do in the morning. We knew that after a good night's sleep, and with daylight, we would be able to work out which way to go.

After dinner, we topped up the other bottles and climbed into our sleeping bags, our heads under the little improvised roof. It almost felt homely as we closed our eyes, exhausted after the day's escapades. The fire burned into the night, keeping us warm and dry. Funnily enough, we slept in until well after nine o'clock in the morning, and after breakfast we were back on our skis with our rucksacks packed. We found the lean-to a few minutes later, about 500m away. Although we were still relatively close to civilisation the night before, without the knowledge of how to manage fire and adapt it to our needs, things could have been very different. I have learnt that it pays to never lose sight of this.

## HOW TO MANAGE FIRE

'Fire is a good servant but a poor master.' This proverb is true; if given boundaries, fire will remain the most obedient of allies. If on the other hand, it is misused or disrespected, it will quickly become an uncaged lion with growing and boundless ferocity – the most destructive of enemies. Fire has no purpose, no aim, no ambition, no mercy; it simply exists and consumes anything that it can in an unbiased fashion. But it is not to be feared, only respected. Tending a fire and using it efficiently means under-standing its traits and habits, and using them to your advantage in order to achieve what you want.

Understandably, in places where people no longer interact with fire on an everyday basis, fire is often misunderstood and seen as a crude tool with only two settings – on or off – but there is far more to it than that. Like an old radio, if handled skilfully, it can be finely tuned to your desire. It can be adapted into anything from a gentle smoulder to a searing column of heat. And it is not only the temperature of a fire that can be adjusted; the shape, size and period of time for which it burns are all variables well within our control.

For temporary visitors to the outdoors, a fire may be called upon regularly for a few main tasks in camp. Wherever you are this list will be similar, although the priorities may change according to the environmental conditions and the nature of your trip. A fire may be used to provide warmth and light, cook food, purify water by boiling, melt snow for drinking water, dry out wet clothes, and keep insects and dangerous game away. This list also applies to the people who call the areas we visit 'home'. Due to their long-term reliance on the bush, though, their fires may be pressed to other jobs too – jobs that are not usually encountered by those just visiting. In many societies, for example, fire is used to dry meat and fish, allowing it to be stored for leaner times. When making traditional items, it is used to change the nature of materials – leaves, wood, stone and bone – or to clear vast swathes of undergrowth. Then there are those rare emergency situations that can be faced by anyone, in which fire can help to attract the attention of rescuers or to cut timber in the absence of cutting tools. And of course, what about the times where we wish to light a fire at home in the fireplace (see OVERLEAF) or in a wood-burning stove?

More than these practical benefits, the most universal and important derivative of being in the presence of a controlled fire is a deep sense of contentedness. Morale is sent skyward when you gaze into bright flames and warm your palms near glowing embers. When I sit with hundreds of miles of trackless wilderness between me and civilisation, it is only my campfire that can bring a sense of home. Without it, communing with the natural world would lose some of its magic and I would leave feeling as if I had not properly visited.

The uses fire can be put to are almost endless, but here I have selected a handful of the most commonly encountered scenarios. The important thing to remember is that fire requires management in order to

| The empty hearth; well swept and cleaned.

|| A layer of screwed-up newspaper tinder.

||| Criss-crossed kindling on top.

|||| Newspaper ignited; kindling takes hold.

||||| Larger fuel added.

|||||| Larger fuel takes hold; the fire reaches full strength.

achieve the desired task in the best way. There is a right way and a wrong way of doing everything, and knowing the correct way can save time, energy, discomfort and disaster.

Except under very rare circumstances, most of the time you will burn wood as your main fuel, as this is usually the most abundant resource available. Other forms of fuel are burned too, and for some people living and travelling in treeless areas are the only alternative.

## FIREWOOD

I cannot overemphasise the importance of collecting dead, dry wood for burning. If you burn green wood that has not dried out, the fire will smoulder and your camp will be swamped in eye-stinging smoke. There are exceptions to this: in the far north of Scandinavia, I have seen the Sami people burn a mix of green and dry birch logs in order to slow down the rate at which the fire consumes fuel. This is only usually attempted in the depths of winter when the moisture content of the wood is at its lowest, and it works very well in these instances. Also in the rainforest, certain species will burn when they are freshly cut. Aside from these examples, most of the time dry wood is burned.

The driest wood will be trees and branches that are still standing because they have not been in contact with the ground. The best of these will be those that are most vertical because rain runs off quickly and does not have a chance to soak in. Having soaked up moisture like a sponge, wood collected from the ground is usually too damp to burn well, so avoid collecting it unless there has been a prolonged warm and sunny spell. If you do decide to collect fallen wood, check the weight of it first. If it feels too heavy for its dimensions, it is likely to be wet inside. It is important to remember firewood has its own character depending on what species of tree it comes from.

As a rule of thumb, soft, light varieties of wood burn fast, producing more flames and therefore more light, while denser woods burn slower, hotter and sometimes with less flame. Light woods tend to burn down completely to ash while harder varieties produce hot, long-lasting embers.

## OTHER FUEL OPTIONS

### DUNG

Where firewood is difficult to find or too expensive, some people form the dung from their livestock into flat 'cakes' and dry them in the sun for a few days before burning them as fuel. In some parts of India, it is common for people to slap these cakes onto the sides of their houses like tiles, removing them once they are dry. Of course, if needed, naturally sun-dried herbivorous dung can also be collected for use.

### OIL SHALE

This occurs in certain areas all around the world and has been used as a fuel in some places for thousands of years. It can be lit with a match, or even crumbled into powder and ignited with sparks from a ferrocerium rod. Again, like dung, this is a fuel very rarely used by wilderness travellers.

### PEAT

Peat, or turf, as it is also known, is a solid fossil fuel formed from ancient ferns, moss and trees which grew in swamps millions of years ago. Traditionally cut into bricks when harvested, it has been used as a fuel for many thousands of years in many different countries and provides about one-third of the heating value of coal. Once harvested, it is pressed to force out water before being stacked up to air-dry.

## COLLECTING AND STORING FUEL

Whatever you use as fuel, make sure you collect enough to last to ensure you don't have to go off on another unexpected search at an inconvenient time. Novices will find it difficult to gauge the quantity needed, but this ability will come with practice. Also, don't waste energy by sawing and splitting wood up any more than necessary.

If it is raining, try to store your fuel in a sheltered spot under a heavily leafed tree or the edge of your tarpaulin. Keep an eye on fuel that is stacked near the fire in case it begins to scorch. If a spot is not available and the fuel must remain out in the rain, lean it up so it stands vertically.

## FIRE SET-UPS

### A QUICK-STOP FIRE

One of the most common reasons we may want to light a fire is to quickly boil the kettle or fry some bacon on a lunch break. If you've been hiking for a few hours through hot, humid jungle or if you've been travelling by snowmobile all morning in minus 30, you can stop when you find an area of dead wood, have a cup of tea and continue with your journey in a matter of minutes. A large fire that burns for hours is not always needed.

In these cases, we need a fire that produces a lot of heat very quickly before burning down to easy-to-deal-with ash. The key is to produce fierce flames with hardly any smoke. You can achieve this by lighting a fire in one of the ways described earlier, and not adding any fuel that is thicker than your little finger. Look for thin twigs, or split a log down into thin splints. Collect plenty of fuel and feed the fire only when needed. Try to burn no more fuel than is necessary. Once the heating is complete and you are enjoying your refreshments, turn your mind towards the clear-up. Do not add any fresh fuel but push any unburned ends into the flames and let it burn down. Once you have finished your break, you should only have cool ash to clear away.

### THE LONG-LOG FIRE

This fire lay is one of the traditional set-ups that old-time hunters and trappers would use when forced to bivouac out in the boreal forest without a sleeping bag. In these cases, it was commonly used in conjunction with an open-fronted lean-to shelter, the shape of which helped to collect the radiated heat and keep it close to the sleeping platform. Depending on the size of logs used, it needs very little tending and will burn for a long period before it needs refuelling. Even if you do have sleeping equipment with you, it remains valid in cold weather when you're not carrying a tent and stove because it allows you and several friends to relax next to it and really warm up. It is also the most suitable set-up for spit-roasting a whole animal, such as a pig. The one thing you do need is timber and plenty of it.

| Look for an area where there are plenty of dead, standing pine trees and make your camp nearby, but not so close that one may fall on you if it were to topple over in high wind.

|| Fell a nice straight one that is at least 25cm thick at waist height and as tall as possible – at least 6m. If they are short trees, fell several.

||| When on the ground, cut it up into 2m-long sections and take them back to camp.

|||| Collect any side branches that snapped off during the felling process. These will be very useful as kindling later.

|||| Once back at camp, trim off and put aside any remaining side branches.

卌 |   Decide exactly where you wish the fire to be, and lay down the two largest logs so they are parallel to each other with a 6cm gap between them. Lay a few small pieces of wood along the ground directly under this gap to form a platform.

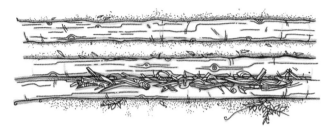

Tinder and kindling laid in place

卌 ||   Now fill the gap by stuffing it with feather sticks or shaving clusters along the whole length, placing any spare shavings or splinters of wood in too. If you have any birch bark, you can add that too.

卌 |||   Snap up a few of the side branches into short lengths and lay them across the gap, like a series of bridges. If there are no side branches, split some lengths out from a spare part of the log.

Sticks laid to
support the third log

卌 ||||   Set fire to the feather sticks all the way along and allow the flames to grow. Once the bridges of wood are burning strongly, place a third 2m-long log on top, so it rests on

208

them. By the time these bridges have burned through, the large logs will be sufficiently alight to burn on their own and will continue to do so with hardly any management. As the logs burn they will reduce in diameter and will become too far apart to maintain a flame. Nudge them together and the flame will soon spring up again.

Established long-log fire with lean-to shelter behind

||||  ||||  Add new logs as and when they are needed, but do so earlier rather than later if you are camping out without a sleeping bag. Adding large new fuel drastically reduces the heat output for a few minutes until it takes hold again.

### CRISS-CROSS FIRE

This arrangement is best if you need to create a deep bed of embers quickly for barbecuing something, or to warm up and dry out a group of people in cold, damp weather.

|     Ignite your tinder, and as the kindling gets going, begin stacking horizontal layers of fuel around it to begin with, and then on top, each layer being set at 90 degrees to the previous one.

||     Gradually increase the thickness of the fuel until it is about wrist-thick. Ensure there is a slight gap in between each piece so as to allow the flames to pass between them and spread upwards. It's easy to get carried away with this stacking process, so don't build it so high that it becomes unsteady and topples over.

INDIAN'S FIRE

I use this arrangement more than any other when I am out in the bush and camping in a different spot each night. It serves the vast majority of needs and requires little effort in setting up or management. The other beauty of this fire is that you don't need to cut up the firewood into lots of neat sections – leave them long and feed them in as they are consumed.

I     Many long lengths of wrist-thick fuel are placed so only their very ends meet in the fire and they radiate out from that central point.

II     When flames are not needed, leave it to burn down by not pushing the ends inwards.

III     When flame is required again, the smouldering ends can be pushed together and it will spring back to life.

## STAR FIRE

This set-up is often confused with the Indian's fire, but there are a couple of slight differences. With the star fire, we use thicker wood – 15cm being ideal. For this reason, it is not the most suitable fire for an overnight camp on the trail because you will probably be left with large, half-burned logs in the morning, which are difficult to clear away. Instead, this is a fire for a longer-term base camp. It works in a very similar way to the Indian's fire, but is vastly more stable: you can safely balance a kettle or other cooking receptacle directly on the logs over the fire without it tipping over.

Providing the weather has been dry, this fire is also very easy to relight because the charred ends begin to smoulder again with only a lick of flame.

|     Use only four lengths of fuel, again keeping them long to minimise the amount of cutting you need to do.

||     When you attend to some chore away from camp, separate the ends slightly – pushing them back together upon your return.

|||     When you wake in the morning, push the charred ends together so they touch, piling up any loose clumps of charcoal around them too.

||||     Now put a bundle of tinder underneath it all and set fire to it. The tinder will give a burst of flame and then die down rapidly, but it will have started the charcoal glowing a little. This can be fanned to encourage it to grow, eventually bursting into flames.

## FINNISH LOG STOVE OR RAAPPANAN FIRE

With the elegant and simple idea of the Finnish log stove, you can take a single log and transform it into a temporary but extremely effective stove for boiling water or cooking. There are several ways of achieving this according to the tools available and your personal preference.

|     First cut a perfectly dry log about 25cm thick and 50cm long. Pine wood is the best because of its resin content, but many other woods will work too.

||     Split the log into three so you end up with two rounded blocks from the exterior and one 4cm-thick board from the middle.

Carving shaving clusters

|||     Stand the two rounded blocks about 4cm apart so the flat inside surfaces face each other.

||||     Now split up the board from the centre into thin splints and carve some shaving clusters.

IIII    Stuff the clusters between the blocks and ignite them. The aim of this short-lived kindling is to get the inside of the blocks burning; once they are, the fire will tick away steadily.

IIII I    A kettle or a frying pan can now be placed on top of the blocks (see OPPOSITE). When you're finished with the stove, pull the blocks apart and tip them over – they will cool quickly and go out in no time.

IIII II    To use the stove again, stand the blocks up again and relight it in a similar fashion.

## ON SNOW

When you need a fire in snowy conditions, you need to light it on the ground and not on the surface of the snow, otherwise it will melt down deeper and deeper, and either go out or provide no benefit and be difficult to use.

I    The answer is to dig down to the ground, ensuring you clear a space large enough to accommodate you and the fire. If you are travelling in the wilderness in these conditions, you should be carrying a means of shifting snow. If circumstances mean you don't have a shovel or a means of improvising, as a last resort you can lay down a platform of long, green branches about as thick as your wrist and light the fire on top of that.

II    Make these branches at least 3m in length; any shorter and the platform will more than likely roll over, fall apart and scatter your fire after a few minutes. The platform will of course eventually burn through, but by that time you will hopefully have established your fire.

## IN STRONG WIND

Although strong wind will help your fire along, fanning it with plenty of oxygen, the unpredictable behaviour of fire in these conditions can make it difficult to work with. The flames sway from one direction to another and back again, which can make even the simplest of tasks, such as boiling some water, much more difficult and time-consuming. You will also find you will burn through your fuel faster. More than all this, though, wind can scatter embers and sparks an impressive distance. Should these land in dry, fibrous undergrowth, things can quickly get out of hand. Do not risk lighting a fire if you're in any doubt; look for a more sheltered spot or create a windbreak.

## PINE KNOTS
### (FOR A BRIGHT FIRE)

When a night-time task requires an extra burst of bright and cheery firelight, a few pine knots placed on the fire will burn for a long time and are perfect for the job. Look for an old standing or fallen pine tree that has begun to decay but is still dry, and try to pull out the small branches from the main trunk. Aim to remove as much of the internal portion of the branch as you can – this is where there is usually a concentration of wood that has become impregnated with wonderful-smelling, flammable resin. It's like pulling a tooth out with the root intact. The wood needs to be slightly soft and crumbly for you to achieve this, but if it is still hard, try tapping the branches with the poll of an axe head or a heavy stick in order to loosen them. Some wind-blown pine stumps also become completely infused with resin, and can be split into splints and burned with a similar effect.

There may be occasions when it will help save valuable time, tinder and kindling if you can keep a fire going for long periods of absence, such as overnight when you are sleeping or during the day when you must head out away from camp for many hours.

In these circumstances, place a short, large-diameter log of dense, hard wood onto a hot bed of embers and leave it to smoulder. If you can, try to place it so one of the end-grain surfaces is in contact with the embers instead of the side of the log – it burns the fibres of the wood like a slow fuse that way. Even if there is a little rain, the log will act as a roof to protect the fire, and there is a good chance it will survive. When you return to it many hours later, turn what is left of the log over, fan it so you can see where the glowing parts are, add some fuel to these areas and continue to fan it. A flame will spring up in no time and you can get the kettle on.

## CARRYING FIRE

For indigenous people all over the world, carrying fire was, and still is, a way of life. They will avoid having to go through the process of lighting a fire whenever possible in order to conserve energy or hard-to-come-by matches. When they move from one place to the next, they may want to quickly light up a cigarette, smoke some bees out when gathering honey, or light a quick cooking fire.

Although Ötzi the iceman never made it to his destination, he has unknowingly recorded one of the ways fire was carried in Europe. In one of the two birch-bark containers he had in his possession at the moment of his death were several fragments of smouldering charcoal wrapped in fresh Norway maple leaves. These embers would have been removed when he arrived at his destination, added to tinder and blown to flames.

One alternative would have been to carry a whole bracket fungus such as *Fomes fomentarius*. Not only can it be made into fine tinder for catching sparks, but it can also be used in its raw state. Having been placed into fire for thirty seconds to set it alight, it will continue to slowly smoulder for hours on its own.

In the remote Bismarck Archipelago of Papua New Guinea, I saw first-hand people carrying fire on several occasions. One man in New Britain showed me how to make a 'slow match' (ABOVE). In the same way that a girl plaits her hair, I watched him plait three strands of coconut husk fibres into a 60cm-long rope that was about as thick as a finger. He offered it to me and I lit it up with a lighter. Once the fibres were burning, he blew the flame out and it continued to smoulder very slowly like a cigar. He told me people would use this method all the time and would sometimes even smuggle lengths of the rope into prison, hiding a burning length in a discreet place so they could light their cigarettes at leisure.

In the neighbouring New Ireland Province, the people have wonderful gardens full of fruit, vegetables, pigs and chickens. These can be vast areas, and tending to them often means walking large distances from their dwellings. Here I encountered them using a special, almost punk wood as a slow match in order to carry their fire with them. They would touch this chunk of wood onto their kitchen fires at home before extinguishing it. This slow match was carried to their gardens, and immediately extinguished once it had been used to kindle a new fire. These are valuable items; they are only burned as long as necessary, and are conserved at all costs.

In the Naga hills of India, the Rengma people made cigar-like rolls of tree bark which would smoulder for use in a similar way.

## FANNING A FIRE

Like the bellows of a blacksmith's furnace, sometimes a fire needs fanning to help it along and speed up the burning. This encouragement is often needed in wet weather. It is like when a car is put into too high a gear for the speed it's doing: it will either stall or will be very sluggish until momentum builds up. If you need to encourage a fire with oxygen, don't burst a blood vessel and make yourself faint by blowing incessantly on it; instead, fan it manually.

You can improvise a fan from any flat, light object with an adequate surface area. Some Native Americans would make beautiful and extremely functional fans from pieces of birch bark sewn together with spruce rootlets. In the tropics, people weave stunningly intricate fans from coconut leaves both for tending their cooking fires as well as cooling themselves. It works so much better and is a lot easier.

Be warned: fanning cannot solve every issue when you have a sluggish fire. Sometimes fanning it will burn out the heart of the fire, which will cause it to die. In these cases, leave it alone to grow by itself.

## HOW TO CLEAR AWAY A FIRE

Whether you are camping overnight close to home or travelling for several weeks deep in the wilderness, the reason you are in that place is probably going to be because you gain enjoyment from being there. One of the greatest experiences for the wilderness traveller is to arrive at a place that has never been visited before (or at least seems that way), and to set up a home for the night. Therefore, we should always consider the enjoyment of like-minded people that may come behind us. Usually this means leaving as little trace of our stay on the landscape as possible. This consideration applies to every chore we attend to in camp.

Nothing we can do has greater potential to leave a mark on the landscape than lighting a fire. We must remember that in lighting one, we take on the responsibility to use it respectfully and with care. It is a great shame when authorities blindly ban people from lighting fires for no real reason but the fear of the mess that may be left behind. In doing so, it takes away the opportunity to learn how to do it properly. When people break the rules anyway, it leads to mess being left out of ignorance or a lack of understanding, rather than through lack of care. Of course, the fire bans that come into force in certain areas and at certain times of year when bushfires are an extreme risk are completely justified; they ensure the preservation of the land we are there to enjoy and the safety of ourselves and others.

In some cases, particularly during extremely cold weather in remote places, it is acceptable and even encouraged to leave a fire scar in place. Being much quicker and easier than establishing a fire from scratch, an old fire with scraps of charcoal and half-burned logs could be a life-saver to someone – perhaps a fisherman who has fallen through thin ice and is in dire need of warming up.

The process of clearing up your fire begins as soon as you have lit it, and it should always be in the back of your mind, guiding your actions from then onwards. Executed well, the whole procedure is a balancing act of estimation as you gauge your fuel requirements according to your needs. Putting a huge log on the fire twenty minutes before you plan to leave is to be avoided. Ideally you will manage your fire so you only have ash and small pieces of charcoal to clear up at the end.

|     Once you know you have finished with your fire, push any half-burned ends into the middle so they burn off.

||     If possible, leave the fire until it has cooled completely before proceeding. If time is of the essence and you need to leave immediately, as is often the case, spread the embers out a little and pour water onto them until they are cool enough to pick up in your hands. Actually feeling the remaining charcoal is the best way to tell if there are any hotspots. Douse everything thoroughly until the remains of the fire are cold.

|||     It is also a good idea to push a sharpened stick into the ground where the fire was to allow some of the water to soak in deeply.

||||     Making sure that they are completely cool, gather up the lumps of charcoal in your hands and scatter them sparsely into the surrounding undergrowth.

IIII     Repeat this until the fire site is completely clean of burned fragments.

Dousing the embers
with water

✚ | Finally, brush some of the material that you cleared before you lit the fire back over the fire site to disguise it and return it back to its original state. If you still have some large pieces of fuel that have been partially burned, ensure they are completely out by dousing them thoroughly or better still, submerging them completely for a few seconds. Tuck them discreetly behind a fallen log or push them into soft, wet mud.

## WOOD-BURNING STOVES
## & OPEN FIREPLACES

One of the greatest pleasures of cold-weather camping has to be relaxing in a tent that is heated with a wood-burning stove. After a hard day pulling a toboggan or driving a snowmobile, they allow us to relax and unwind in an environment that could otherwise offer discomfort and threat. A friend and I once tested the ability of a two-man canvas tent combined with the small, lightweight stove we were using and had some surprising results. It was minus 40°C outside, but in our tent the thermometer read plus 28°C. The same goes for stoves or open fires at home; they bring a cheer to the room that is impossible to replicate.

Every stove and fireplace has its own personality and quirks, and it takes a while to get used to one, but once you do it becomes a reliable friend. It is important to take the time to learn how to manage them, both to get the best performance from them, and also to avoid danger. If you follow the advice in this book, you will not go far wrong when it comes to performance, as many of the techniques for building a fire can be adapted for use in these enclosed hearths.

## THE DANGERS OF CARBON MONOXIDE

Carbon monoxide is a very dangerous and potentially lethal gas that is produced when wood or other carbon compounds are burned without an adequate oxygen supply. Make doubly sure that whenever you have a fire of any kind in an enclosed space such as a tent or shelter that it is properly ventilated with a good supply of air to promote complete combustion. There have been several cases of people perishing in this way; in the early 2010s, a young girl died after a disposable barbeque was left inside the tent in which she was camping.

Carbon monoxide is a silent killer, being both colourless and odourless, so it is vitally important that we remain alert, and look out for the signs and symptoms of poisoning in ourselves and our companions. The effects of such poisoning include headache, lethargy and general weakness, difficulty concentrating, and nausea. As time goes on the headache can become far worse and the skin can become flushed red. The tiredness caused by this poisoning frequently results in victims falling asleep, meaning they become unable to aid themselves. I have spent countless nights sleeping out next to a wood-burning stove without any issues, and there can be no better accompaniment to falling asleep than hearing the gentle ticking of a stove and seeing the shadows dancing on the canvas walls. But do ensure you have proper ventilation. Keep an eye on yourself and others, and you will come to no harm.

## ABOUT THE AUTHOR

Daniel Hume is an instructor at Woodlore, Ray Mears' School of Wilderness Bushcraft, and is an expert on bushcraft, tracking and survival in the wild. The art of making and keeping a fire has been Daniel's greatest passion from a young age. Daniel has made it his mission to travel the world and learn first-hand how tribesmen in remote places build fires key to their survival and existence. In his quest he has mastered some of the most extraordinary native fire-making techniques from across the globe.

# ACKNOWLEDGEMENTS

In life I believe it is the people around me that make it possible to channel my vision and turn it into reality. This project has been no different and I am indebted to so many.

Firstly I would like to express the deepest thanks to my parents John and Deborah for their love, support and encouragement – I simply couldn't have done this without them.

To my girlfriend Angelica Garcia who has not only assisted with the photography and translation but has been by my side throughout this journey. Thank you for your love, patience and advice.

To Oliver Rednall, an honest man who tells it straight – one of the most valuable attributes to have in a friend.

To my friend and companion on many a trip, Rob Bashford, one of the finest and most professional of outdoor leaders.

To Ray Davis, one of the last true Suffolk countrymen, for demonstrating for the camera.

Special thanks goes to the many people around the world that I have had the good fortune to meet. In my quest for fire they have gone from strangers to friends and have been selfless with their generosity, hospitality and wisdom. For me, you have proven that the world is a good place even when you lift its veil and dare to venture to the untouched corners that often receive bad press.

Deepest gratitude goes to Kunibert Tabil of New Ireland Province, Papua New Guinea, for believing in my project and introducing me to the fire plough and those that still depend upon it. Also to Pak Martin Makusi and family as well as Moses of Gam Island, West Papua, for welcoming me to their home and for demonstrating the hand drill and bamboo strike-a-light. Big thanks to Jonas Wenda of Wamena, Papua, for being the most professional of guides. Also to the many other members of remote communities living in the bush that have helped me - it has been a privilege to meet and learn from you.

A heartfelt thank you to John Perry who first put me in touch with Furniss Lawton. His simple but invaluable advice - 'Just sit down and type' - really got the ball rolling.

To Eugenie Furniss and Rory Scarfe for listening to my rough idea and seeing the potential value in it. To Rory especially, thank you for all your tireless work, advice and commitment to the project. Thanks also to the rest of the team at Furniss Lawton - Liane-Louise Smith, Rachel Mills, Isha Karki and Lucy Steeds - for successfully putting the book out into the wider world and for their help.

Massive thanks to Georgina Rodgers who has not only helped me put my knowledge and experiences down on paper, but has been a beacon of light and a true guiding star through this whole process.

Huge thanks to the team at Century, particularly Ajda Vučićević, my editor, and Becky Millar. It's been a pleasure to work with such a positive and passionate group of people.

Thank you to the hugely talented Adam Doughty - whose fantastic illustrations have not only added clarity to the instructional text but also a uniquely special touch.

To Tim Barnes for working his magic and ensuring the design and layout are imbued with the spirit of the subject.

To Woodlore and its many students for the fun times and the amazing opportunities I have been given - I am honoured and privileged to be a part of the team. Sincere thanks to Ray Mears for inspiring me, showing me how to travel in wild places and guiding me along the right path for many years.

**Daniel Hume**